T0259636

DFT-Domain Based Single-Microphone Noise Reduction for Speech Enhancement

A Survey of the State-of-the-Art

Synthesis Lectures on Speech and Audio Processing

DFT-Domain Based Single-Microphone Noise Reduction for Speech Enhancement: A Survey of the State-of-the-Art
Richard C. Hendriks, Timo Gerkmann, and Jesper Jensen
2013

Speech Recognition Algorithms Using Weighted Finite-State Transducers
Takaaki Hori and Atsushi Nakamura
2013

Articulatory Speech Synthesis from the Fluid Dynamics of the Vocal Apparatus
Stephen Levinson, Don Davis, Scot Slimon, and Jun Huang
2012

A Perspective on Single-Channel Frequency-Domain Speech Enhancement
Jacob Benesty and Yiteng Huang
2011

Speech Enhancement in the Karhunen-Loève Expansion Domain
Jacob Benesty, Jingdong Chen, and Yiteng Huang
2011

Sparse Adaptive Filters for Echo Cancellation
Constantin Paleologu, Jacob Benesty, and Silviu Ciochina
2010

Multi-Pitch Estimation
Mads Græsbøll Christensen and Andreas Jakobsson
2009

Discriminative Learning for Speech Recognition: Theory and Practice
Xiaodong He and Li Deng
2008

Latent Semantic Mapping: Principles & Applications
Jerome R. Bellegarda
2007

Dynamic Speech Models: Theory, Algorithms, and Applications
Li Deng
2006

Articulation and Intelligibility
Jont B. Allen
2005

DFT-Domain Based Single-Microphone Noise Reduction for Speech Enhancement: A Survey of the State-of-the-Art

Richard C. Hendriks, Timo Gerkmann, and Jesper Jensen

ISBN: 78-3-031-01436-9 paperback
ISBN: 78-3-031-02564-8 ebook

DOI 10.1007/978-3-031-02564-8

A Publication in the Springer series
SYNTHESIS LECTURES ON SPEECH AND AUDIO PROCESSING
Lecture #11
Series Editor: B.H. Juang, *Georgia Tech*
Series ISSN
Synthesis Lectures on Speech and Audio Processing
Print 1932-121X Electronic 1932-1678

DFT-Domain Based Single-Microphone Noise Reduction for Speech Enhancement

A Survey of the State-of-the-Art

Richard C. Hendriks
Delft University of Technology, The Netherlands

Timo Gerkmann
Universität Oldenburg, Germany

Jesper Jensen
Oticon A/S, Denmark
Aalborg University, Denmark

SYNTHESIS LECTURES ON SPEECH AND AUDIO PROCESSING #11

ABSTRACT

As speech processing devices like mobile phones, voice controlled devices, and hearing aids have increased in popularity, people expect them to work anywhere and at any time without user intervention. However, the presence of acoustical disturbances limits the use of these applications, degrades their performance, or causes the user difficulties in understanding the conversation or appreciating the device. A common way to reduce the effects of such disturbances is through the use of single-microphone noise reduction algorithms for speech enhancement.

The field of single-microphone noise reduction for speech enhancement comprises a history of more than 30 years of research. In this survey, we wish to demonstrate the significant advances that have been made during the last decade in the field of discrete Fourier transform domain-based single-channel noise reduction for speech enhancement. Furthermore, our goal is to provide a concise description of a state-of-the-art speech enhancement system, and demonstrate the relative importance of the various building blocks of such a system. This allows the non-expert DSP practitioner to judge the relevance of each building block and to implement a close-to-optimal enhancement system for the particular application at hand.

KEYWORDS

single-microphone, speech enhancement, noise reduction

Contents

Acknowledgments

The authors wish to acknowledge Regin Kopp Pedersen, Oticon A/S for providing graphical material on the "invisible-in-the-ear canal" hearing aid.

Richard C. Hendriks, Timo Gerkmann, and Jesper Jensen
January 2013

Glossary

AMS — analysis-modification-synthesis
AR — auto-regressive
CDFT — complex-discrete Fourier transform
DAM — diagnostic acceptability measure
DD — decision-directed
DCT — discrete cosine transform
DFT — discrete Fourier transform
FFT — fast Fourier transform
GARCH — generalized autoregressive conditional heteroscedasticity
HINT — hearing in noise test
HMM — hidden Markov model
IDFT — inverse discrete Fourier transform
IIC — invisible-in-the-canal
IMCRA — improved minima controlled recursive averaging
KLT — Karhunen-Loeve transform
MAP — maximum a posteriori
MDFT — magnitude-discrete Fourier transform
ML — maximum likelihood
MMSE — minimum mean-square error
MOS — mean opinion score
MS — mininum statistics
MSE — mean-square error
MVDR — minimum-variance distortion-less response
PDF — probability density function
PESQ — perceptual evaluation of speech quality
PSD — power spectral density
SEG-SNR — segmental SNR
SNR — signal-to-noise ratio
SPP — speech presence probability
SRT — speech reception threshold

STFT	short-time Fourier transform
STOI	short-time objective intelligibility
STSA	short-time spectral amplitude
TCS	temporal cepstrum smoothing
VAD	voice activity detector

CHAPTER 1

Introduction

During the last decade there has been an increasing interest in mobile speech processing applications such as voice controlled devices, mobile telephony, smart phone applications, hearing aids, etc. As these applications have gained popularity, users expect them to work anywhere and at any time. This, obviously, imposes heavy demands on the robustness of these devices. Specifically, the user's environment may present acoustical disturbances like passing cars, trains, competing speakers, office noises, etc., in addition to the target (speech) source of interest. The impact of these disturbances can be severe, for example for mobile telephony and voice controlled devices, but also for the hearing aid case, where limitations of an impaired auditory system prevent the user from separating target speech from disturbance.

A popular solution is to process the noisy signal to reduce the impact of the disturbances and enhance the sound quality, i.e., listener comfort, or speech intelligibility. This can generally be done using single-microphone or multi-microphone noise reduction methods. Although multi-microphone methods often lead to better performance than single-microphone methods, the additional costs, e.g., for matched microphones, power usage, computational complexity, and size demands, sometimes limit their usability. A relevant application where only a single-microphone solution is possible, is the "invisible-in-the-ear canal" hearing aid, see Figure 1.1(a)–Figure 1.1(d). This battery-driven device is small enough to be inserted into the ear canal, and to be essentially invisible. The location and the size of the device prevents the use of multiple microphones, hence, the device is equipped with a single microphone. The difficulties that hearing aid users face, particularly in noisy environments, make a noise reduction algorithm an essential part of the signal processing in the device. Although single-channel noise reduction algorithms typically only provide very modest improvements in intelligibility [1, 2], they do improve quality aspects of the signals [3], which helps to increase the comfort and reduce listener's fatigue.

Single-channel noise reduction techniques are also important in the context of multi-microphone systems, because these can often be decomposed into a concatenation of a beamformer, i.e., a spatial filter, and a single-channel noise reduction algorithm (Figure 1.1(e)). In a sense, beamformers and single-channel methods are complementary: the beamformer performs a spatial filtering, reducing noise sources originating from other directions than the target, and possibly adapting to the spatial changes in the acoustical scene. The single-microphone method, on the other hand, aims at reducing the remaining noise originating from the same direction as the target, and typically adapts to the *temporal* changes of the statistics of the sound sources involved. In fact, it can be shown, under assumptions which are often approximately valid, that an optimal strategy for multi-microphone noise reduction is to apply a minimum-variance distortion-less response (MVDR) multi-microphone beamformer and process the output signal by a single-microphone noise reduction algorithm [4, 5]. Thus, single-channel noise reduction systems play an important role in many applications, either as stand-alone systems, or as a post-processing step for multi-microphone methods.

After more than 30 years of research on the topic of single-microphone noise reduction for speech enhancement, this research field may be considered mature. Although the research community in this field is still extremely active, the field may seem to advance rather slowly. Is this because the field has converged to a point beyond which hardly any improvements can be obtained? Are there no big challenges left? Or are there important challenges, which are difficult to formulate in a mathematically tractable way?

The goals of this book are threefold. First, we wish to demonstrate that significant progress has in fact been made over the last decade, perhaps particularly in terms of extending the range of applications where single-channel speech enhancement algorithms can be employed successfully. Secondly, our goal is to provide a concise description of the various building blocks in a state-of-the-art speech enhancement system, allowing a non-specialist to implement such a system. Finally, we wish to demonstrate through simulation experiments, the relative importance of the various building blocks, allowing the DSP practitioner to judge the relevance of the building blocks for a particular application.

The book is structured as follows. In Sec. 2 we present the problem of single-channel speech enhancement, and discuss the general methodologies that exist for solving the problem. Sec. 3 limits our focus to a class of single-channel enhancement systems which may be called discrete Fourier transform (DFT)-based enhancement systems, and presents general assumptions made in this class of algorithms. Sec. 4–Sec. 7 discuss four key building blocks in a state-of-the-art DFT-based speech enhancement system, namely, algorithms for estimating speech DFT coefficients based on their noisy observations, algorithms for estimating the probability that speech is present in a noisy DFT coefficient observation, algorithms for estimating the noise power spectral density, and, finally, algorithms for estimating the clean speech power spectral density. Sec. 8 outlines methodologies for objective and subjective evaluation of speech enhancement systems. Sec. 9 provides a simulation study of various speech enhancement systems of increasing complexity, allowing the reader to appreciate the impact of various building blocks on speech enhancement performance. Finally, in Sec. 10 we summarize the challenges which still remain in the area of single-channel speech enhancement and outline possible future paths to approach these challenges.

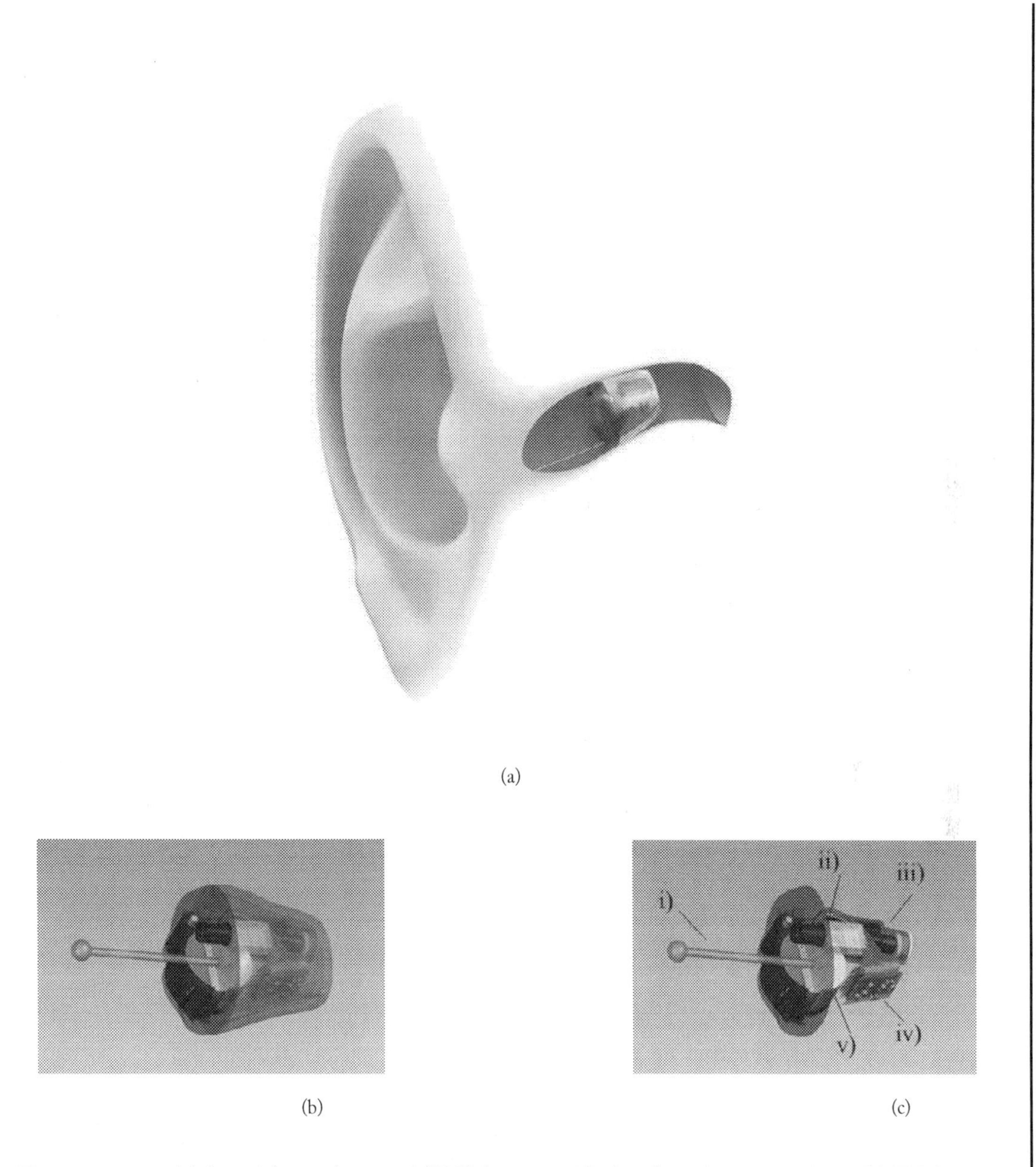

Figure 1.1: 1.1(a) Invisible-in-the-canal (IIC) hearing aid placed in the ear canal. 1.1(b) IIC hearing aid, transparent shell. 1.1(c) interior of IIC hearing aid. i) pull-out string, ii) microphone, iii) loudspeaker, iv) amplifier and integrated circuit (including signal processor), v) battery.

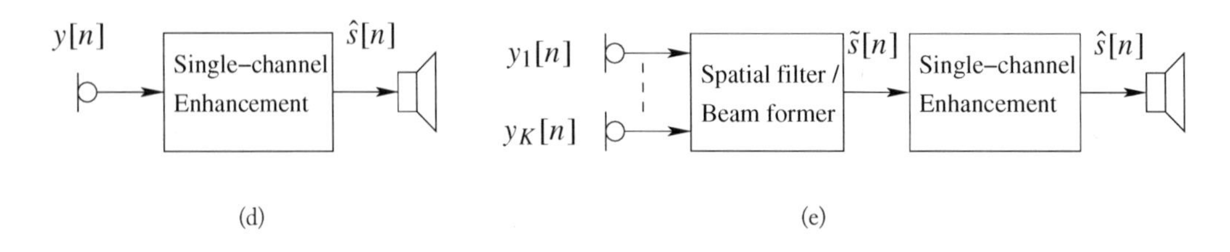

Figure 1.1: 1.1(d) Block diagram of single-channel enhancement system. 1.1(e) Single-channel enhancement system as post-processor for multi-channel system.

CHAPTER 2

Single Channel Speech Enhancement–General Principles

With an active research history of more than 30 years, the single-channel speech enhancement area is vast, and an exhaustive historical coverage is infeasible. Instead, our goal in this section is to give an overview of the general ideas and principles behind the most successful single-channel speech enhancement systems.

We generally consider an additive signal model of the form

$$y[n] = s[n] + n[n], \tag{2.1}$$

where $y[n]$, $s[n]$, and $n[n]$, represent the observed noisy signal, the unknown target signal, and the noise signal, respectively, and where n is a discrete-time index. Most often, we consider the signals as realizations of stochastic processes, and assume that the target process and the noise process are statistically independent. The additive and independent noise model underpins most recent speech enhancement research and is valid in a wide range of practical situations; even in the case of convolutional noise, e.g., speech signals contaminated by room reverberance, this additive model has successfully been used in enhancement systems to suppress the negative impact of late reflections (see, e.g., [6]). With this statistical signal model in mind, the speech enhancement problem is often posed as an estimation problem: given a realization $y[n]$ of the noisy process, find an estimate $\widehat{s}[n]$ of the target realization $s[n]$. What distinguishes speech enhancement systems is i) the domain in which the estimation is performed, e.g., time, frequency, subspace, wavelet, etc., ii) the types of prior assumptions made about the target and noise process, and iii) the distortion measure chosen to quantify the quality of the estimate.

Despite the many varieties of single-channel enhancement systems, it turns out that the basic tasks to be solved are rather universal. Figure 2.1 depicts a general block diagram for solving these tasks; most existing systems consist of some or all of the blocks in this figure.

2.1 ANALYSIS-MODIFICATION-SYNTHESIS (AMS) SYSTEM

The top row of blocks in Figure 2.1 shows an analysis-modification-synthesis (AMS) system, which, based on a noisy input speech signal $y[n]$,[1] produces the estimate $\hat{s}[n]$ of the underlying clean speech signal $s[n]$. The AMS system divides the noisy input signal into generally overlapping signal frames $\underline{y}_\ell$ within which underlying statistical properties can be assumed invariant. The frames are windowed by an analysis window, before, typically, a transform is applied to the windowed frame to produce transform coefficients $y_k(\ell)$ where ℓ and k are the frame and transform coefficient index, respectively. The noisy transform coefficients are then modified, often by applying a scalar gain to each coefficient independently, before the enhanced transform coefficients are inversely transformed, and the resulting time frames $\hat{\underline{s}}_\ell$ are overlap-added using some synthesis window, to produce the enhanced noise reduced signal $\hat{s}[n]$.

The transform serves several purposes. First, it may act as a decorrelator to deliver noisy transform coefficients which are uncorrelated or even statistically independent. In this case, it turns out, optimality in many distortion criteria can be obtained by processing the noisy transform coefficients independently. The class of subspace methods, e.g., [7, 8, 9, 10, 11] aim at achieving exactly that by applying a Karhunen-Loeve transform (KLT) to the noisy signal frames. Although theoretically optimal, applying a KLT at this stage is computationally costly as the transform is signal dependent and must be re-computed for each noisy signal frame; furthermore, the resulting transform coefficients are not linked straightforwardly to physical properties, such as frequency content of the incoming signals. For this reason, most existing systems compute short-time Fourier transforms (STFTs) using a discrete Fourier transform (DFT), e.g., [12, 13, 14, 15, 16]. The DFT can be implemented computationally efficient, it delivers approximately uncorrelated transform coefficients, in practice it leads to enhancement performance similar to that of the theoretically better justified subspace methods, and it allows a straightforward interpretation in terms of spectral signal content. The latter point is important because some systems are based on models of speech production, e.g., [17] or auditory models, e.g., [18], which are both most naturally described in the Fourier domain rather than, e.g., the KLT domain (although, as we shall see below, subspace methods which operate with auditory models do exist). Other transform options exist, such as the discrete cosine transform (DCT), e.g., [19], or wavelet transforms, e.g., [20].

Some systems are formulated in the time domain. For example, the Hidden Markov Models (HMMs)-based systems introduced by Ephraim [21] form maximum *a posteriori* (MAP) or minimum mean-square error (MMSE) estimates of target frames by left-multiplying Toeplitz-structured Wiener filter matrices onto the noisy frame vector $\underline{y}_\ell$, to implement a convolution operation. It should be clear, though, that this operation can also be interpreted as processing transform coefficients independently, e.g., by using as transform the orthogonal matrix with eigen vectors of the Wiener filter matrix as columns.

[1] We tacitly ignore analog-to-digital (A/D) and digital-to-analog (D/A) conversion, and simply use the discrete-time index n for all signals involved.

When the DFT and the inverse DFT (IDFT) transforms are used, the signal chain at the top of Figure 2.1 can be interpreted as a decimated analysis-modification-synthesis filter bank, where the analysis filterbank is derived using the analysis window as a prototype filter, the decimation rate corresponds to the number of samples between successive analysis windows, and the synthesis filter bank uses as prototype filter the synthesis windows employed in the overlap-add block.

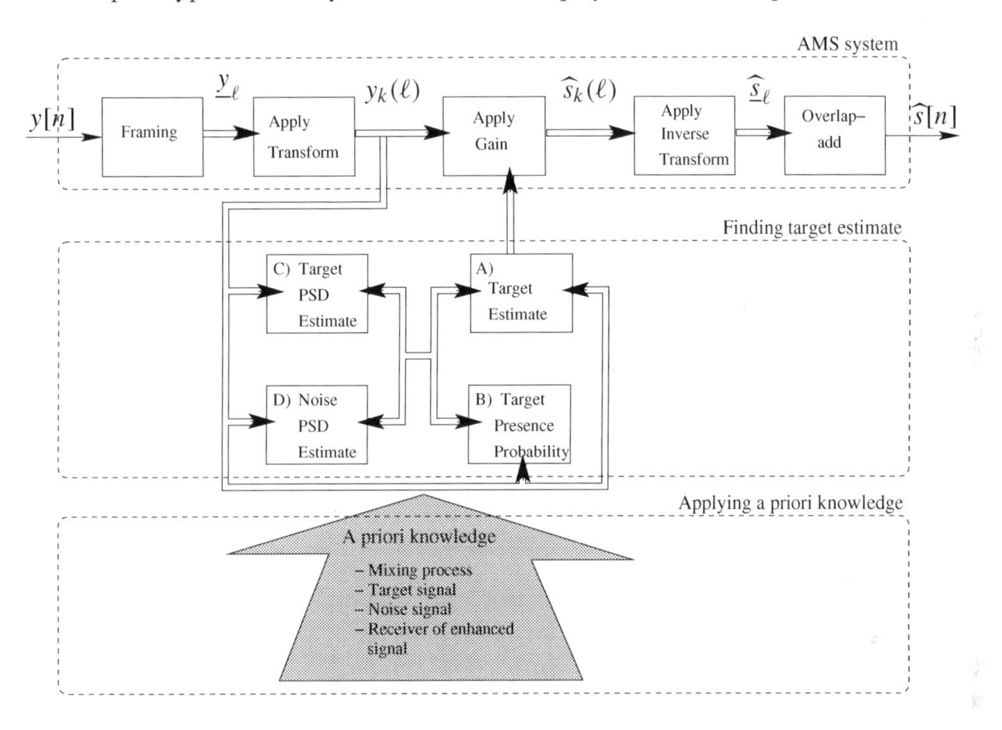

Figure 2.1: General single-channel speech enhancement system. Thin arrows indicate sequences of time domain samples, thick arrows indicate sequences of (processed) signal frames, i.e., vector-valued signals.

2.2 FINDING THE TARGET ESTIMATE

The AMS chain described above is simply a system that allows appropriate processing of a noisy input signal, often by applying gain values to the noisy transform coefficients, to produce an enhanced output signal. The critical point in any enhancement system is how to find the gain values that lead to a suitably enhanced output signal using only the noisy input signal and potentially prior knowledge and assumptions with respect to the acoustical situation.

The subblock "A) Target Estimate" computes these gain values; typically gain values are found which are optimal in some statistical sense, e.g., MMSE. Often, estimates of the target signal are derived under the assumption that target and noise are both present in the noisy observation. However, this is not always the case since speech signals have pauses between syllables and words, and, furthermore, many speech transform coefficients are essentially zero, even during speech activity. Thus, optimal estimators take into account this speech presence uncertainty, e.g., [22], by estimating the speech presence probability from the noisy observation, hence the subblock "B) Target Presence Probability Estimate." It turns out that the optimal gain values for most proposed criteria are functions of the short-time power spectral densities of the target and noise signals. None of these are available in isolation, and must therefore be estimated from the noisy signal. Hence, the subblocks "C) Target power spectral density (PSD) Estimation" and "D) Noise PSD Estimation." We will treat the subblocks A)–D) in much more detail in the following sections.

2.3 *A PRIORI* KNOWLEDGE AND ASSUMPTIONS

In order to estimate a target signal from a noisy observation, one must have some knowledge about the signals in question, for example that noise is additive and independent of the target signal. Early, heuristically motivated, methods exist, which are based on these assumptions only. For example, the spectral subtraction approach [12, 13] is a method for finding an estimate of the target STFT, which will be discussed in Sec. 4.2. The spectral subtraction approach is computationally extremely simple, but is often not optimal in any sense, and is hardly ever used as a stand-alone algorithm, as it tends to produce an annoying musical-like residual noise.

Generally speaking, any available *a priori* knowledge could (and should) be used in the estimation process. For single-channel speech enhancement, this *a priori* knowledge comes in three different classes, namely, i) knowledge about the target signal, ii) knowledge about the noise signal, and iii) the knowledge (or assumption) that the enhanced signal is to be listened to by a human listener. These classes of knowledge are clearly not mutually exclusive; in fact, enhancement systems exist which utilize knowledge from all three categories. However, while systems which rely heavily on *a priori* knowledge or assumptions may perform very well when the assumptions are valid, they may perform poorly when the acoustical situation does not match the underlying assumptions. Thus, exploiting *a priori* knowledge involves a tradeoff between maximum system performance and robustness to changing acoustical scenarios.

2.3.1 TAKING SPEECH SIGNAL CHARACTERISTICS INTO ACCOUNT

Speech signals have certain characteristics due to the fact that they have been produced by the human speech production system. For DFT- or KLT-based systems, it has been shown that the transform coefficients tend to be super-Gaussian distributed[2] [19, 24]. This observation, combined with the assumption that transform coefficients are statistically independent across time and coefficient index, has been used in numerous contributions for deriving various estimators of target transform coefficients (subblock A), e.g., [11, 15, 25, 26, 27, 28].

[2] This observation is approximately valid at least for the KLT, DFT, and discrete cosine transform (DCT). The exact distribution depends on the frame length used [19] and on the actual speech sound [23], and on the way in which the target PSD is estimated in the enhancement system.

The assumption that target coefficients are statistically independent across time is mainly made for convenience, since with a typical delay between successive analysis frames in the order of 15 ms, and with a frame duration in the order of 30 ms, speech DFT coefficients are surely interdependent [16]. Cohen [16] has developed several estimators taking into account this dependency. Perhaps most notably, statistically consistent estimators were proposed for the speech PSD, as alternatives to the so-called decision-directed estimator of Ephraim and Malah [14]; the decision-directed estimator is more heuristically motivated, but often used due to its simplicity and rather good performance.

Also due to the speech production process, speech spectra tend to show dependencies across frequency. The fact that the speech power spectral envelope tends to follow that of an auto-regressive (AR) model was perhaps first used by Lim and Oppenheim [17] who developed a system for iteratively finding the MAP estimate of the target AR PSD, which was then used in a Wiener-based enhancement system. However, the sequences of AR spectra obtained by the Lim and Oppenheim system [17] were recognized to not always be realistic for natural speech [29]. Thus, Hansen and Clements [29] improved the method by constraining the sequence to have certain speech like characteristics, while Sreenivas and Kirnapure [30] proposed a codebook-based system, which constrained the method to use realistic speech spectra stored in a codebook. More recently, cepstral methods have been used to exploit the structure of the speech spectral envelope. For example, Breithaupt et al. [31] estimated the spectral envelope of the target PSD in the cepstral domain, exploiting the fact that the spectral envelope is coded in a small set of the cepstral coefficients.

The general idea of constructing more realistic spectro-temporal models of the signals involved was further developed by Ephraim who proposed an HMM to model jointly the target signal and the noise signal as time-varying AR processes [21] (see also e.g., [32] for an example of newer systems which rely on the same philosophy). Target and noise PSD sequences were represented jointly by state sequences, and the optimal state sequence (and the enhanced frame sequence) could, e.g., be found via Viterbi decoding. Building on the idea of taking the spectro-temporal dependencies of speech DFT coefficients into account, Cohen observed that large magnitude DFTs tend to follow large magnitudes and small magnitudes tend to follow small magnitudes, and proposed to model this behavior using generalized autoregressive conditional heteroscedasticity (GARCH) models [33]. Methods for estimating the speech PSD based on GARCH models were proposed in [34].

More recently, the class of corpus or inventory driven systems [35, 36] have been proposed; rather than modeling the temporal dynamics of the target speech signal using statistical models such as HMMs or GARCH models, this class of methods stores a database of speech material, and tries to estimate the underlying clean target frames by matching the sequences of observed noisy frames to sequences of clean frames in the clean speech corpus. While these systems can deliver excellent performance, they are sensitive to mismatches between the signals used to form the corpus or train the HMM and the observed signals, and, furthermore, they tend to have significant storage and computation demands which make them unsuited for some applications.

2.3.2 TAKING NOISE PROCESS CHARACTERISTICS INTO ACCOUNT

The noise to be suppressed by an enhancement system is application specific, and characterizing it in any precise manner is difficult. Often, the noise process, and consequently, the noise DFT coefficients, have been assumed to be Gaussian distributed; observing that noise sources are often a sum of several independent sources, the central limit theorem ensures that the observed noise process tends to be Gaussian. This assumption is underlying most estimators of target DFT coefficients, e.g., [14, 15, 37].

Existing methods for estimating the noise PSD from the noisy signal rely on the assumption that the noise PSD evolves "more slowly" than the speech production process, such that a noise PSD estimate obtained in the past (say 100 ms ago) is still valid at the current time. The minimum statistics (MS) algorithm of Martin updates the noise PSD estimate in minima of the trajectory of the noisy PSD [38], assuming that speech is absent in these time-frequency regions.

More recently, methods have been proposed which estimate the target signal-to-noise ratio or the target presence probability to decide to which extent a given time-frequency point can be used for updating the noise PSD estimate, e.g., [39, 40, 41, 42]; particularly for non-stationary noise sources these methods can achieve better performance than the MS method.

While the MS method and its derivatives exploit temporal characteristics of the noise process, the same principle can be applied to the spectral characteristics. More specifically, the PSD of noisy voiced speech often decays to the noise PSD level between the harmonics of the speech signal; assuming then that the noise PSD is locally constant across frequency, the noise PSD in spectral regions where the harmonics are dominant can be extrapolated from the estimates obtained between the harmonics [43, 44].

2.3.3 TAKING THE HUMAN AUDITORY SYSTEM INTO ACCOUNT

Enhancement systems often produce an enhanced signal by minimizing a distortion measure, e.g., short-time spectral amplitude (STSA) mean-square error (MSE) $\mathrm{E}(|A_k(\ell) - \widehat{A}_k(\ell)|^2)$, where $A_k(\ell)$ and $\widehat{A}_k(\ell)$ are random variables representing the target and estimated spectral amplitude, respectively. Ideally, the choice of distortion measure reflects the perceptual aspects of interest, e.g., the speech quality or intelligibility. Yet, one often adheres to simple MSE-based distortion measures for mathematical convenience, and simply because the distortion measure which optimally reflects speech quality or intelligibility is not known. In an attempt to improve the perceptual quality of the enhanced signals, many alternatives to the STSA MSE have been proposed. For example, shortly after proposing the STSA MSE, Ephraim et al. proposed estimators for the log-spectral amplitude MSE [37] to be better in line with the compressive characteristics of the healthy cochlea. Furthermore, Loizou et al. [45] derived estimators for several perceptually motivated distortion measures, and You et al. [46] proposed the β-order MMSE spectral amplitude estimator. Thus, the choice of distortion measure is a trade-off between perceptual relevance and mathematical tractability. The estimator proposed in [47, 48] by Breithaupt et al., generalizes β-order MMSE optimal spectral amplitude estimation for parameterizable super-Gaussian speech priors.

Other, more explicit attempts to take into account the human auditory system in the enhancement process include methods based on simple models of the auditory system. The dominant idea here is to apply masking models, e.g., [49] (which was originally developed in the area of audio and speech coding to reduce the perceptual impact of quantization noise) to estimate a masking threshold for each noisy frame; the masking threshold expresses in the spectral domain the amount of additive, uncorrelated noise which can be added to the underlying clean target signal and yet remain undetectable [3]. Noting that the enhancement process often introduces audible artifacts, Tsoukalas et al. [50] proposed to use the masking threshold to identify spectral regions where the noise is audible and apply the processing only in the audible noise regions. Other methods rely on the fact that the estimation error signal, i.e., the difference between the underlying clean target and the enhanced target, can be decomposed into a sum of two uncorrelated components [9]: the target distortion, i.e., the distortion introduced by the enhancement process to the underlying target signal, and the residual noise, i.e., the part of the original noise signal remaining in the enhanced signal. Generally, these undesired components cannot both be eliminated, and a trade-off exists. Several systems have been proposed which aim at minimizing the signal distortion power while keeping the PSD of the residual noise below the masking threshold (e.g., [10, 51]), but other reasonable possibilities of using the masking threshold exist, e.g., [18, 52, 53]. Generally speaking, this class of methods tends to produce enhanced signals with somewhat fewer artifacts than conventional methods, while their noise reduction capabilities are roughly similar.

While the masking effects exploited in the systems above are relatively well-understood and can largely be explained by processes in the auditory periphery, methods do exist which aim at exploiting the processing believed to take place at higher stages of the auditory pathway. For example, Paliwal [54] has proposed to restore the modulation spectrum of the underlying clean speech signal, that is the frequency content of the time series consisting of short-time Fourier transform magnitudes. It is well known that modulations are represented in the neural signal reaching the auditory cortex [55] and that modulation frequencies in the range 4–20 Hz are important for the intelligibility of speech, e.g., [54, 56, 57, 58]. The algorithm in [54] shows promising results, and the full potential of this type of method remains to be seen.

[3] Observe, though, that in order to compute this masking threshold, an initial estimate of the target PSD is usually needed.

CHAPTER 3

DFT-Based Speech Enhancement Methods–Signal Model and Notation

We now focus in more detail on the very large class of single-channel speech enhancement methods called short-time Fourier transform (STFT)-based speech enhancement methods, i.e., we use the DFT as the transform in Figure 2.1. The specific enhancement systems which we consider are rather general, as they impose relatively few assumptions concerning the speech and noise production process; thus, the resulting systems are robust and work well in diverse acoustical situations. Furthermore, the resulting systems tend to be relatively simple in terms of computational and storage complexity, which makes them of interest in a wide range of possible applications, including mobile communication devices, hearing aids, head sets, etc.

We generally apply the additive-and-independent noise model in Eq. (2.1), and we consider all signals as realizations of underlying stochastic processes. Thus, signal frames are realizations of vector random variables, and transform coefficients are realizations of scalar random variables. We represent random variables by capital letters and realizations thereof as lowercase letters. Thus, the statistical model for DFT coefficients at frequency index k and frame index ℓ, is given by

$$Y_k(\ell) = S_k(\ell) + N_k(\ell) \tag{3.1}$$

where $Y_k(\ell)$, $S_k(\ell)$, and $N_k(\ell)$ are zero-mean (complex-valued) random variables representing DFT coefficients of the noisy observation, the speech target, and an additive noise term, respectively. For mathematical convenience, we use the standard assumptions that $S_k(\ell)$, and $N_k(\ell)$ are statistically independent and that signal frames are sufficiently long and that their overlap is sufficiently small, such that DFT coefficients are approximately independent across time and frequency [59]; for this reason, without loss of generality, we may drop time and frequency indices, and simply write $Y = S + N$. Let $R = |Y|$, $A = |S|$, and $W = |N|$ denote random variables representing the noisy, clean, and noise spectral magnitude, respectively. Furthermore, we introduce the spectral variances, $\sigma_S^2 = E(A^2)$ and $\sigma_N^2 = E(W^2)$, and let $\xi = \sigma_S^2/\sigma_N^2$ and $\zeta = r^2/\sigma_N^2$ denote the *a priori* and *a posteriori* SNR [22], respectively. Notice that ζ is in fact a realization (through r) of the random variable $Z = R^2/\sigma_N^2$.

To model the observation that speech spectral magnitudes are sparsely distributed, see e.g., [24], we assume that speech DFT magnitudes $A \geq 0$ are distributed according to a probability density function (PDF) of the form

$$p_A(a;\gamma,\nu) = \frac{\gamma\beta^{\nu}}{\Gamma(\nu)} a^{\gamma\nu-1} \exp\left(-\beta a^{\gamma}\right), \ \gamma > 0, \nu > 0, a \geq 0, \tag{3.2}$$

where $\Gamma(\cdot)$ is the Gamma function. For given parameters γ, ν, and spectral variance σ_S^2, the parameter $\beta > 0$ is fully determined. For example, for $\gamma = 1, \beta = \sqrt{\nu(\nu+1)/\sigma_S^2}$ and for $\gamma = 2, \beta = \nu/\sigma_S^2$ [28]. Note that for $\gamma = 2$ the generalized Gamma function can be written in the same mathematical form as the χ-distribution, which has been used in [27, 47, 48] to derive MMSE estimators. Other special cases of Eq. (3.2) are the Exponential distribution for $\gamma = 1, \nu = 1$ and the Rayleigh distribution for $\gamma = 2, \nu = 1$. In Figure 4.1(d) examples are shown for $p_A(a, \gamma = 1, \nu = 0.6)$, $p_A(a, \gamma = 1, \nu = 1.5)$ and $p_A(a, \gamma = 2, \nu = 1)$ where $p_A(a, \gamma = 1, \nu = 0.6)$ provides a good model of speech DFT magnitude distributions in the present context [27, 28, 48]. The corresponding phase variable is assumed to be uniformly distributed in $[0; 2\pi)$, and independent of A. Finally, we generally assume that the noise DFT coefficients N follow a zero-mean (complex) Gaussian distribution with variance σ_N^2. For use in the following chapters, we note that the conditional density $p_{Y|S}(y|S = s)$ therefore is complex Gaussian with mean s and variance σ_N^2, that is,

$$p_{Y|S}(y|S = s) = \left(\pi\sigma_N^2\right)^{-1} \exp\left(-\frac{1}{\sigma_N^2}|Y - s|^2\right). \tag{3.3}$$

CHAPTER 4

Speech DFT Estimators

As we focus on the class of STFT-based enhancement methods and use the DFT, the goal is to produce an estimate of the noise-free complex speech DFT coefficient S at each time-frequency point. Historically, two different estimator classes have been developed, namely, complex-DFT (CDFT) estimators which estimate the complex-valued DFT coefficient S directly, and magnitude-DFT (MDFT) estimators which estimate $A = |S|$, and append the noisy phase to form a complex DFT coefficient. Assuming a uniformly distributed phase that is independent of the magnitude, using the noisy phase can be shown to be optimal in the MMSE sense, e.g., [14, 60]. MDFT estimators were originally proposed instead of CDFT estimators, as the speech phase was thought to be perceptually less important (see e.g., [14, 17, 22]). More recently, it has been argued that the role of phase in speech enhancement has been underestimated in the past [61, 62]. Krawczyk and Gerkmann [63] have proposed an algorithm to blindly estimate the clean speech phase in voiced speech from a noisy observation. They showed that a blind estimation of the clean speech phase is possible and may push the limits of speech enhancement algorithms further. The benefits of having an estimate of the clean speech phase at hand are twofold: first, it can be appended to the estimated magnitude. Secondly, the information in the clean speech phase estimate can be directly employed in the estimation of the clean speech spectral amplitudes, and serve as additional means to distinguish noise outliers from speech [64].

4.1 STATISTICAL MODELING ASSUMPTIONS

The derivation of DFT estimators is subject to statistical assumptions which we now discuss via speech DFT-coefficient histograms. Estimation of speech DFT histograms is non-trivial as speech is a non-stationary process. Although several methods have been proposed for this purpose, the discussion on how to accurately measure speech DFT histograms is still on-going (e.g., [65], [66]). Here we measure the histograms as in [15, 24], where the decision-directed approach [14] (see Sec. 7 for details) is used in order to collect speech DFT realizations with a similar variance. In this way the obtained histograms are conditioned on the estimated speech PSD, and changing the used speech PSD estimator could in principle change the histogram.

Figure 4.1(a) shows the joint histogram of real and imaginary parts of speech DFT coefficients, denoted by $S_{\Re}$ and $S_{\Im}$, respectively, estimated from roughly 5.5 hours of speech from the Timit database [67]. From this histogram several conclusions can be drawn. First, the histogram is circular symmetric, which implies the phase to be uniformly distributed. Secondly, for any A, the probability of the phase Φ remains the same, i.e., Φ and A are independent, that is, $p_{A,\Phi}(a, \phi) = p_A(a)\,\frac{1}{2\pi}$.

Thirdly, real and imaginary parts of speech DFT coefficients are uncorrelated as $p_{S_\Im|S_\Re}(s_\Im)$ is symmetric around $s_\Im = 0$, and therefore,

$$\mathrm{E}(S_\Re S_\Im) = \int_{-\infty}^{\infty} s_\Re \, p_{S_\Re}(s_\Re) \int_{-\infty}^{\infty} s_\Im \, p_{S_\Im|S_\Re}(s_\Im) \, ds_\Im ds_\Re = 0. \tag{4.1}$$

Finally, even though the real and imaginary parts are uncorrelated, they are *not* independent. This can be verified by measuring the marginal histograms of real and imaginary parts, i.e., $p_{S_\Re}(s_\Re)$ and $p_{S_\Im}(s_\Im)$ and verifying that $p_{S_\Re,S_\Im}(s_\Re, s_\Im) \neq p_{S_\Re}(s_\Re)\, p_{S_\Im}(s_\Im)$. The product of the marginal histograms is shown in Figure 4.1(b). Clearly, Figure 4.1(a) and Figure 4.1(b) are not identical, and thus real and imaginary parts of DFT coefficients are not independent.

Another aspect is the shape of the speech DFT coefficient PDF. Figure 4.1(c) shows a cross-section of the histogram in Figure 4.1(a) for $s_\Im = 0$. In addition, a Gaussian distribution with the same variance is shown. Clearly, the speech DFT histogram is more peaked and more heavy-tailed than the Gaussian distribution, i.e., it tends to be super-Gaussian. Figure 4.1(d) shows the histogram of speech MDFT coefficients based on the same data as in Figure 4.1(a) together with three example distributions that can be obtained from the PDF in Eq. (3.2). Specifically, $p_A(a, \gamma = 1, \nu = 0.6)$ and $p_A(a, \gamma = 1, \nu = 1.5)$ leading to super-Gaussian and $p_A(a, \gamma = 2, \nu = 1)$ leading to complex-Gaussian distributed CDFT coefficients, respectively. The super-Gaussian distribution shows a better fit to the peak and tail of the MDFT histogram and can therefore be expected to lead to better speech DFT estimators than when the CDFT coefficients are assumed to be complex-Gaussian distributed.

4.2 SPECTRAL SUBTRACTION

The earliest methods of noise reduction were based on so-called power spectral subtraction. One way to derive this estimator is based on the maximum likelihood (ML) estimate of the speech PSD [22], given the PDF $p_Y(y; \sigma_S^2, \sigma_N^2)$ of the noisy DFT coefficients. Assuming that speech and noise DFT coefficients are complex-Gaussian distributed, the ML estimate of σ_S^2 is given by, $\widehat{\sigma_S^2} = |y|^2 - \sigma_N^2$, explaining the name: power spectral subtraction. Taking the square-root of $\widehat{\sigma_S^2}$ and appending the noisy phase, then leads to $\widehat{s} = gy$, with

$$g = \left(1 - \frac{\sigma_N^2}{|y|^2}\right)^{\frac{1}{2}}. \tag{4.2}$$

Many modifications have been proposed in order to improve its quality (see e.g., [12, 13, 17, 22]). One of these is the use of a half-wave rectifier to guarantee that $\widehat{\sigma_S^2}$ is positive. Another important modification is to use smoothed values of $|y|^2$ across time in order to reduce errors in the estimate $\widehat{\sigma_S^2}$.

Generally, spectral subtraction type-of estimators tend to be somewhat heuristically motivated. For instance, the derivation of power spectral subtraction is based on an ML estimate of the speech

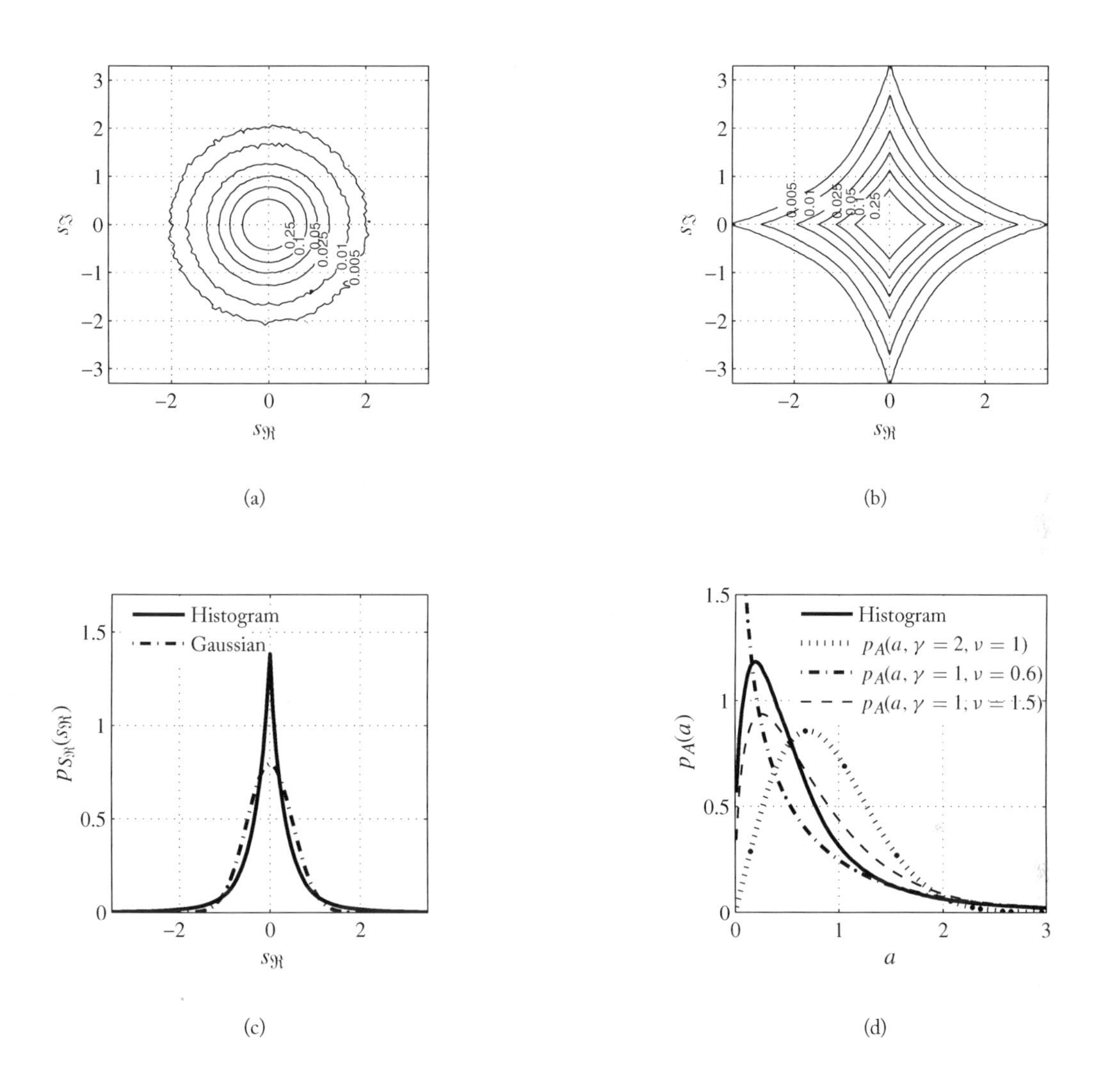

Figure 4.1: (a) Joint histogram of CDFT coefficients. (b) Product of the marginal histograms of real and imaginary parts of DFT coefficients. (c) Cross-section of joint histogram from Figure 4.1(a) along the line $s_{\Im} = 0$. (d) Histogram of MDFT coefficients.

PSD instead of an estimate of the speech DFT S or its magnitude A. Further, the estimator uses no direct prior information of the PDFs of S and N, but only indirect information via the PDF of Y.

4.3 LINEAR MMSE ESTIMATORS

A theoretically more solid approach is to derive estimators of the underlying clean DFT coefficient which are optimal in some statistical sense. In the remainder of this chapter, we focus primarily on estimators which are optimal in the MMSE sense. While other estimators, e.g., ML or MAP estimators, tend to lead to simpler expressions, our experience is that MMSE-based estimators lead to slightly better performance. The perhaps simplest approach is to estimate the clean speech DFT coefficients using a *linear* MMSE estimator, often referred to as the frequency domain Wiener filter [17, 68]. Although both linear CDFT and MDFT estimators [69] exist, the linear CDFT MMSE estimator has obtained most attention. The linear CDFT MMSE estimator can be derived by minimizing the MSE while constraining the estimator to be linear in Y, that is, $\widehat{S} = gY$ with $g = \arg\min_g \mathrm{E}(|S - gY|^2)$. Using complex derivatives [70], this leads to $\widehat{S} = \sigma_S^2(\sigma_S^2 + \sigma_N^2)^{-1}Y$. The linear complex-DFT MMSE estimator thus only depends on the speech and noise PSD, and is independent of the actual distribution of speech and noise DFT coefficients. It can be shown that if the speech and noise DFT coefficients are complex Gaussian, then the frequency domain Wiener filter is optimal amongst all estimators. Generally speaking though, *non-linear* MMSE estimators do quite a better job.

Although the linear MDFT MMSE estimator is less well known in the field of speech enhancement than its complex counterpart, it offers some potential advantages. The linear MDFT MMSE estimator is derived by setting $\widehat{A} = gR + m$, where g and m can be found by joint minimization of $\mathrm{E}\big((A - \widehat{A})^2\big)$ with respect to g and m, leading to [69]

$$\widehat{A} = \max\left(\frac{\mathrm{E}(AR) - \mathrm{E}(R)\mathrm{E}(A)}{\mathrm{E}(R^2) - \mathrm{E}(R)^2}\,(R - \mathrm{E}(R)) + \mathrm{E}(A), 0\right), \tag{4.3}$$

where the max operator is used to take into account that magnitudes are positive by definition. Given the distributional assumptions made in Sec. 3, the expressions $\mathrm{E}(A)$, $\mathrm{E}(R)$, and $\mathrm{E}(AR)$ can be computed in terms of σ_S^2, σ_N^2 and the distributional parameter ν of the density in Eq. (3.2) [69].

While the linear CDFT estimator only depends on σ_S^2 and σ_N^2 and is independent of the assumed distribution of speech and noise DFT coefficients, the linear MDFT estimator does depend on the assumed distribution of the speech MDFT coefficients by means of the ν-parameter, as the relation between $\mathrm{E}(A^2)$ and $\mathrm{E}(A)$, $\mathrm{E}(R)$, and $\mathrm{E}(AR)$ is distribution dependent. By properly adapting the ν-parameter such that the PDF in Eq. (3.2) matches with the super-Gaussian histogram in Figure 4.1(d), the linear MDFT MMSE estimator in Eq. (4.3) can be shown to clearly improve in terms of segmental signal-to-noise ratio (SNR) improvement over a linear MDFT MMSE estimator based on a non-super-Gaussian distribution.

4.4 NON-LINEAR MMSE ESTIMATORS

Non-linear estimators [71] offer more potential than their linear counterparts, partly because the speech DFT coefficients tend to be non-Gaussian, and partly because prior knowledge on the speech and noise processes can be taken into account by means of the speech and noise PDF. Non-linear estimators can be derived in a Bayesian framework by minimizing a non-negative cost-function. Depending on whether the goal is to estimate the CDFT or the MDFT coefficients, the expected cost-function is given by $\mathcal{R} = \mathrm{E}\big(\mathcal{C}(S, \hat{S})\big)$ or $\mathcal{R} = \mathrm{E}\big(\mathcal{C}(A, \hat{A})\big)$, respectively. Often, the squared-error cost function is used, e.g., [14, 15, 28, 72, 73], as it is mathematically tractable and often leads to good performance. Minimizing the squared error cost-function leads to the MMSE or conditional mean estimator [71], i.e., $\hat{s} = \mathrm{E}(S|Y)$ or $\hat{a} = \mathrm{E}(A|Y)$, for the CDFT and the MDFT coefficients, respectively. Other cost-functions have been proposed that may be perceptually more relevant (see e.g., [45, 46, 47, 48]). Usually, these cost-functions lead to estimators that can be formulated as variants of conditional mean estimators or can be computed using similar mathematics. Without claiming to be complete, we give in Tables 4.1 and 4.2 an overview of MDFT and CDFT estimators, respectively, as presented in literature with their cost-function and distributional assumptions.

MMSE estimators can in general be written as an *a priori* and *a posteriori* SNR dependent gain function, say $g(\xi, \zeta)$, applied to the noisy DFT-coefficient. That is, $\hat{s} = g_s(\xi, \zeta)y$ or $\hat{a} = g_a(\xi, \zeta)y$, for CDFT and MDFT estimators, respectively. The *a priori* SNR ξ is generally unknown and must be estimated from the noisy data, as will be discussed in Sec. 7.

Already in 1984, Ephraim and Malah [14] presented an MMSE magnitude-DFT estimator under the assumptions that both speech and noise CDFT coefficients have a complex-Gaussian distribution. However, as demonstrated in Figure 4.1(c), histograms of speech DFT coefficients tend to show a super-Gaussian PDF, suggesting that better estimators could be found if these statistical characteristics were taken into account. Initiated by Martin, CDFT MMSE (e.g., [15, 73, 74] [28]), MDFT MMSE (e.g., [27, 28, 75]), and MDFT MAP estimators (e.g., [24, 27, 75]) have been presented under a super-Gaussian PDF for the speech DFT coefficients.

The development of super-Gaussian CDFT and MDFT estimators as outlined above was made under conflicting assumptions. Specifically, to derive the CDFT MMSE estimators in [15, 28], it was assumed that the real and imaginary parts of DFT-coefficients are independent. However, this is clearly not in line with measured histograms shown in Figure 4.1. Further, it was assumed that the real and imaginary parts have a double-sided generalized-Gamma distribution, which is a double-sided version of the PDF in Eq. (3.2) (or one of its special cases), given by

$$p_{S_\Re}(s_\Re; \gamma, \nu) = \frac{\gamma \beta^\nu}{2\Gamma(\nu)} |s_\Re|^{\gamma\nu - 1} \exp\left(-\beta |s_\Re|^\gamma\right), \ \gamma > 0, \nu > 0. \tag{4.4}$$

Assuming a variance of $\mathrm{var}(S_\Re) = \sigma_S^2/2$, special cases of Eq. (4.4) are the Gaussian distribution for $\gamma = 2$, $\nu = 0.5$, $\beta = 1/\sigma_S^2$, the Laplace distribution for $\gamma = 1$, $\nu = 1$, $\beta = 2/\sigma_S$, and the double sided Gamma distribution [76, Ch. 5.11] for $\gamma = 1$, $\nu = 0.5$, $\beta = \sqrt{3/2}/\sigma_S$. In contrast, for the MDFT MMSE estimators, e.g., [28], it was assumed that the speech phase Φ is uniformly distributed and the magnitude A is generalized-Gamma distributed. While the statistical assumptions made to derive the MDFT estimators are in line with the histograms in Figure 4.1, there is no consistency between the statistical models for the CDFT and the MDFT estimators.

The problem was solved in [60, 77] where a unified framework was presented that allows to derive both MDFT and CDFT estimators under the same statistical assumptions, that are in line with the histograms in Sec. 4.1.

In order to estimate the MDFT coefficients $\hat{a} = g_a(\xi, \zeta)r$, one can compute the gain function

$$g_a = \mathrm{E}(A|y)r^{-1} = \frac{1}{r}\frac{\int_0^{+\infty}\int_{-\pi}^{+\pi} a\, p_{Y|S}(y|a,\phi)\; p_A(a)\; p_\Phi(\phi)\, d\phi da}{\int_0^{+\infty}\int_{-\pi}^{+\pi} p_{Y|S}(y|a,\phi)\; p_A(a)\; p_\Phi(\phi)\, d\phi da}, \tag{4.5}$$

by substituting the PDF in Eq. (3.2) for $p_A(a)$ and the PDF in Eq. (3.3) for $p_{Y|S}(y|a,\phi)$ [28].

To compute the gain-function $g_s(\xi, \zeta)$ for the CDFT coefficients one can compute

$$g_s = \mathrm{E}(S|y)y^{-1} = \frac{e^{-j\theta}}{r}\frac{\int_{-\infty}^{+\infty}\int_{-\infty}^{+\infty}(s_\Re + js_\Im)\, p_{Y|S}(y|s_\Re, s_\Im)\; p_S(s_\Re, s_\Im)\, ds_\Re ds_\Im}{\int_{-\infty}^{+\infty}\int_{-\infty}^{+\infty} p_{Y|S}(y|s_\Re, s_\Im)\; p_S(s_\Re, s_\Im)\, ds_\Re ds_\Im}, \tag{4.6}$$

where θ is the phase of y. In [15,28], Eq. (4.6) was computed by assuming that the real and imaginary parts of complex speech DFT coefficients are independent. A better alternative is to rewrite Eq. (4.6) using polar-transformations, as proposed in [60], leading to

$$g_s = \mathrm{E}(S|y)y^{-1} = \frac{1}{r}\frac{\int_0^{+\infty}\int_{-\pi}^{+\pi} a e^{j(\phi-\theta)}\; p_{Y|S}(y|a,\phi)\; p_A(a)\; p_\Phi(\phi)\, d\phi da}{\int_0^{+\infty}\int_{-\pi}^{+\pi} p_{Y|S}(y|a,\phi)\; p_A(a)\; p_\Phi(\phi)\, d\phi da}. \tag{4.7}$$

Based on the distributions in Eq. (3.2) and Eq. (3.3), Eq. (4.7) can be computed according to the correct statistical modeling assumptions in Sec. 4.1. These assumptions are consistent with the statistical assumptions made to derive the MDFT estimators. In addition, using Eq. (4.7) and Eq. (4.5) it can be shown [60] that the gain function g_s is always real and positive, and, $g_s \leq g_a$, from which it follows that the CDFT MMSE estimators always lead to similar or more suppression than the MDFT MMSE estimators.

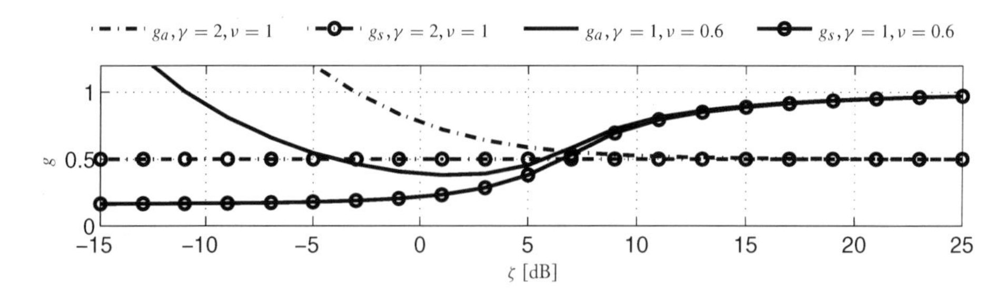

Figure 4.2: Gain curves g_a and g_s for magnitude-DFT and complex-DFT estimators for $\xi = 0$ dB.

Figure 4.2 shows the gain curves g_a and g_s for an *a priori* SNR of $\xi = 0$ dB derived under the assumption that the CDFT coefficients are complex-Gaussian distributed, i.e., using $\gamma = 2$

Table 4.1: Overview of magnitude-DFT estimators presented in literature.

Magnitude-DFT Estimators			
Cost function	Estimator $\hat{A}$	Assumed Distribution of A	ref.
$\mathrm{E}\left((A-\hat{A})^2\right)$	$\mathrm{E}(A\|y)$	Eq. (3.2), $\gamma=2$, $\nu=1$	[14]
$\mathrm{E}\left((A-\hat{A})^2\right)$	$\mathrm{E}(A\|y)$	Eq. (3.2), $\gamma=1$ and $\gamma=2$	[27, 28, 78]
$\mathrm{E}\left((A^2-\widehat{A^2})^2\right)$	$\mathrm{E}(A^2\|y)$	Eq. (3.2), $\gamma=2$, $\nu=1$	[72]
$\mathrm{E}\left((A^2-\widehat{A^2})^2\right)$	$\mathrm{E}(A^2\|y)$	Eq. (3.2), $\gamma=2$	[79]
$\mathrm{E}\left((\log\{A\}-\log\{\hat{A}\})^2\right)$	$\exp\{\mathrm{E}(\log\{A\}\|y)\}$	Eq. (3.2), $\gamma=2$, $\nu=1$	[37]
$\mathrm{E}\left((\log\{A\}-\log\{\hat{A}\})^2\right)$	$\exp\{\mathrm{E}(\log\{A\}\|y)\}$	Eq. (3.2), $\gamma=2$	[80]
$\mathrm{E}\left((A^\beta-\hat{A}^\beta)^2\right)$	$\mathrm{E}(A^\beta\|y)^{1/\beta}$	Eq. (3.2), $\gamma=2$, $\nu=1$	[46]
$\mathrm{E}\left((A^\beta-\hat{A}^\beta)^2\right)$	$\mathrm{E}(A^\beta\|y)^{1/\beta}$	Eq. (3.2), $\gamma=2$	[47, 48]
$\mathrm{E}\left(A^p(A-\hat{A})^2\right)$	$\frac{\mathrm{E}(A^{p+1}\|y)}{\mathrm{E}(A^p\|y)}$	Eq. (3.2), $\gamma=2$, $\nu=1$	[45]
$\mathrm{E}\left(A^{-2p}(A^\beta-\hat{A}^\beta)^2\right)$	$\left(\frac{\mathrm{E}(A^{\beta-2p}\|y)}{\mathrm{E}(A^{-2p}\|y)}\right)^{1/\beta}$	Eq. (3.2), $\gamma=2$, $\nu=1$	[81]

and $\nu = 1$ in Eq. (3.2), and under the assumption that the CDFT coefficients are super-Gaussian distributed, i.e., using $\gamma = 1$ and $\nu = 0.6$ in Eq. (3.2). It is clear that under the same distribution, $g_s \leq g_a$, and the estimators derived under the super-Gaussian distribution are more sensitive to the *a posteriori* SNR ζ.

Generally, the speech quality of these super-Gaussian-based estimators improves over estimators which assume complex-Gaussian distributed DFT coefficients (see e.g., [28]). As mentioned above and as shown in Figure 4.2, the super-Gaussian-based estimators are functions of the *a priori* and *a posteriori* SNR. Given an estimate of the noise PSD (see Sec. 6), the *a posteriori* SNR is given, but the *a priori* SNR is unknown and has to be estimated. This may lead to under- or over-estimates and thus to over- or under-suppression of the noise. An additional advantage of the super-Gaussian-based estimators is that they are generally non-linear and show more dependence on the realization based *a posteriori* SNR ζ (see Figure 4.2). This can help to compensate under- or over-estimates of the *a priori* SNR to some extent.

Table 4.2: Overview of complex-DFT estimators presented in literature.

Complex-DFT Estimators			
Cost function	Estimator $\hat{S}$	Assumed Distribution of S	ref.
$\mathrm{E}\left((S-\hat{S})^2\right)$	$\frac{\xi}{\xi+1}y$	Eq. (4.4), $\gamma=2, \nu=0.5$	[68]
$\mathrm{E}\left((S_\Re-\hat{S}_\Re)^2\right)$	$\mathrm{E}(S_\Re \mid y_\Re)$	$p_S(s)=p_{S_\Re}(s_\Re)\ p_{S_\Im}(s_\Im)$, Eq. (4.4), $\gamma=1, \nu=0.5$	[15, 73]
$\mathrm{E}\left((S_\Re-\hat{S}_\Re)^2\right)$	$\mathrm{E}(S_\Re \mid y_\Re)$	$p_S(s)=p_{S_\Re}(s_\Re)\ p_{S_\Im}(s_\Im)$, Eq. (4.4), $\gamma=1, \nu=1$	[15, 74, 82]
$\mathrm{E}\left((S_\Re-\hat{S}_\Re)^2\right)$	$\mathrm{E}(S_\Re \mid y_\Re)$	$p_S(s)=p_{S_\Re}(s_\Re)\ p_{S_\Im}(s_\Im)$, Eq. (4.4), $\gamma=1$ and $\gamma=2$	[28, 83]
$\mathrm{E}\left((S-\hat{S})^2\right)$	$\mathrm{E}(S \mid y)$	Eq. (3.2), $\gamma=1$ and $\gamma=2$, with $p_\Phi(\phi)=\frac{1}{2\pi}$	[60, 77]

CHAPTER 5

Speech Presence Probability Estimation

We will now turn our attention to the speech presence probability (SPP) estimation [block B) in Figure 2.1]. We define two hypotheses: $\mathcal{H}_{1,k}(\ell)$ indicates that speech is present in frequency bin k at time segment ℓ while $\mathcal{H}_{0,k}(\ell)$ indicates speech absence. In the sequel we neglect the time and frequency indices for brevity. In Figure 2.1 we see that the speech presence probability (SPP) may be needed for the target estimate, the noise PSD estimate, and the speech PSD estimate. For instance, if we knew at which time-frequency point speech were absent, we could use these time-frequency points to update the noise power spectral density estimate (see Sec. 6). Also, statistically optimal estimators for the clean speech spectral coefficients (Sec. 4.4) are derived under the assumption that speech is actually present [14, 84]. The optimal estimator under speech presence uncertainty is thus given as

$$\begin{aligned} \mathrm{E}(C(S) \mid y) &= P(\mathcal{H}_1 \mid y)\, \mathrm{E}(C(S) \mid y, \mathcal{H}_1) + P(\mathcal{H}_0 \mid y)\, \mathrm{E}(C(S) \mid y, \mathcal{H}_0) &(5.1)\\ &= P(\mathcal{H}_1 \mid y)\, \mathrm{E}(C(S) \mid y, \mathcal{H}_1), &(5.2) \end{aligned}$$

where $C(S)$ is a function of the clean speech DFT coefficients, such as the magnitude $C(S) = A$, $P(\mathcal{H}_1 \mid y)$ is the *a posteriori* SPP, $\mathrm{E}(C(S)|Y, \mathcal{H}_0) = 0$ [85] and $\mathrm{E}(C(S) \mid Y, \mathcal{H}_1)$ is realized by the estimators in Sec. 4.4. For a discussion of the special case $C(S) = \log(A)$ see also [84]. In this section we aim at deriving estimators for the probability $P(\mathcal{H}_1 \mid y)$ that speech is present at a given time-frequency point. The probability $P(\mathcal{H}_0 \mid y)$ that speech is absent then follows from $P(\mathcal{H}_0 \mid y) = 1 - P(\mathcal{H}_1 \mid y)$.

5.1 *A POSTERIORI* SPEECH PRESENCE PROBABILITY

The *a posteriori* SPP is given as the probability that hypothesis $\mathcal{H}_1$ is true given a realization y of the noisy observation $Y = |Y|\mathrm{e}^{\mathrm{j}\Theta}$. While the distribution of $|Y|$ changes whether it is conditioned on $\mathcal{H}_0$ or $\mathcal{H}_1$, we assume that the distribution of the phase Θ is independent of $\mathcal{H}_1$. Thus, we may write

$$P\left(\mathcal{H}_1 \mid y, \sigma_N^2, \sigma_S^2\right) = P\left(\mathcal{H}_1 \mid |y|, \sigma_N^2, \sigma_S^2\right) = P\left(\mathcal{H}_1 \mid \zeta, \sigma_N^2, \sigma_S^2\right), \tag{5.3}$$

where the *a posteriori* SNR $\zeta = |y|^2/\sigma_N^2$ is a realization of the random variable $Z = |Y|^2/\sigma_N^2$. Conditioning the probability on ζ instead of y has the advantage that we can use parameterizable distributions for the spectral amplitudes, similar to the distributional assumption used to derive estimators of the clean speech spectral amplitudes in Sec. 4.

For brevity we will drop the condition on the speech and noise PSDs σ_S^2 and σ_N^2. Using Bayes's theorem, for the *a posteriori* SPP we have

$$P(\mathcal{H}_1 \mid \zeta) = \frac{P(\mathcal{H}_1)\, p_{Z|\mathcal{H}_1}(\zeta)}{P(\mathcal{H}_0)\, p_{Z|\mathcal{H}_0}(\zeta) + P(\mathcal{H}_1)\, p_{Z|\mathcal{H}_1}(\zeta)}, \tag{5.4}$$

where $P(\mathcal{H}_0)$ and $P(\mathcal{H}_1)$ denote the *a priori* speech absence and speech presence probabilities, respectively. In Eq. (5.4), the likelihoods of speech presence and absence are compared for a given realization of the *a posteriori* SNR ζ. If the likelihood of speech presence is much larger than the likelihood of speech absence, Eq. (5.4) results in

$$P(\mathcal{H}_1 \mid \zeta)\Big|_{p_{Z|\mathcal{H}_1}(\zeta) \gg p_{Z|\mathcal{H}_0}(\zeta)} = 1. \tag{5.5}$$

Thus, in Eq. (5.4) we compare the given observation ζ to two models, one for speech absence and one for speech presence. These models are given by the likelihood functions $p_{Z|\mathcal{H}_0}(\zeta)$ and $p_{Z|\mathcal{H}_1}(\zeta)$. The values of the likelihood functions for a given realization ζ indicate how well the realization fits the respective model, and the *a posteriori* SPP $P(\mathcal{H}_1 \mid \zeta)$ indicates the probability that $p_{Z|\mathcal{H}_1}(\zeta)$ is the better model for the time-frequency point under consideration. If the model for speech absence yields a much better fit than the model for speech presence, we obtain

$$P(\mathcal{H}_1 \mid \zeta)\Big|_{p_{Z|\mathcal{H}_1}(\zeta) \ll p_{Z|\mathcal{H}_0}(\zeta)} = 0. \tag{5.6}$$

To compute the *a posteriori* SPP, models are needed for the *a priori* probabilities $P(\mathcal{H}_0)$ and $P(\mathcal{H}_1) = 1 - P(\mathcal{H}_0)$, as well as the likelihood functions for speech presence $p_{Z|\mathcal{H}_1}(\zeta)$ and speech absence $p_{Z|\mathcal{H}_0}(\zeta)$. In the SPP literature, most commonly Gaussian distributions are assumed for the real and imaginary part of the noisy observation [14, 84, 85]. As a result, the PDF of the noisy periodogram and the *a posteriori* SNR is given by an exponential distribution [76]. More flexibility is given, if the χ^2-distribution is assumed as proposed in [86]. Note that if spectral amplitudes follow a χ-distribution, the resulting *a posteriori* SNR is χ^2-distributed. The χ-distribution is a special case of Eq. (3.2) with shape parameter ν, $\gamma = 2$, $\beta = \nu/\sigma_S^2$. The χ^2-distribution can be written in the identical mathematical form as a Gamma-distribution. As in speech absence we have $\mathrm{E}(\zeta) = 1$, it follows for the likelihood of speech absence that

$$p_{Z|\mathcal{H}_0}(\zeta) = \frac{1}{\Gamma(\nu)} \nu^{\nu}\, \zeta^{\nu-1} \exp(-\nu\, \zeta)\,. \tag{5.7}$$

The likelihood under speech presence is given as [86]

$$p_{Z|\mathcal{H}_1}(\zeta) = \frac{1}{\Gamma(\nu)} \left(\frac{\nu}{1+\xi_{\mathcal{H}_1}}\right)^{\nu} \zeta^{\nu-1} \exp\!\left(-\nu\, \frac{\zeta}{1+\xi_{\mathcal{H}_1}}\right), \tag{5.8}$$

with the model parameter $\xi_{\mathcal{H}_1}$, which reflects the *a priori* SNR that is expected when speech is present. For the *a posteriori* SPP it then follows that

$$P(\mathcal{H}_1 \mid \zeta) = \left(1 + \frac{P(\mathcal{H}_0)}{P(\mathcal{H}_1)}(1 + \xi_{\mathcal{H}_1})^{\nu} \exp\left(-\nu\zeta \frac{\xi_{\mathcal{H}_1}}{1 + \xi_{\mathcal{H}_1}}\right)\right)^{-1}. \tag{5.9}$$

With the Gamma distribution, different shapes of the PDFs can be formed using the parameter ν. For $\nu = 1$ we obtain the exponential distribution for the PDFs and the same expression for the *a posteriori* SPP that is found for a Gaussian modeling of the complex noisy observation, e.g. [14, 84, 85]. The shape parameter ν can be increased to model the distribution of a smoothed observation and will result in a steeper transition between the large and small SPPs for a given ζ [86]. Rather different expressions for the *a posteriori* SPP result if different distributional assumptions are made for the clean speech and noise coefficients, e.g., a Gaussian model for noise, and a super-Gaussian or parameterizable model for speech [15], [48], [87].

While in most publications on SPP estimation the distributional assumptions are rather similar (exponential [14, 84, 85] or Gamma [86, 88]), the difference between algorithms lies primarily in the estimation of the model parameter $\xi_{\mathcal{H}_1}$ and the priors $P(\mathcal{H}_0)$, $P(\mathcal{H}_1)$.

5.2 ESTIMATION OF THE MODEL PARAMETER $\xi_{\mathcal{H}_1}$

5.2.1 SHORT-TERM ADAPTIVE ESTIMATE

Often, the model parameter $\xi_{\mathcal{H}_1}$ is adaptively estimated on a short time scale using the decision-directed (DD) approach [14] (also see Sec. 7.1), e.g., in [14, 15, 84, 85] or based on power subtraction [89]. However, these short-term estimators aim at estimating the SNR ξ present in the current time-frequency point. As a consequence, the likelihood function of speech presence would actually model the PDF of Z for the time-frequency point under consideration. This can be a problem when we aim at detecting speech absence: in speech absence, the short-term estimates yield SNR estimates close to zero, such that $\xi_{\mathcal{H}_1} \approx 0$ and $p_{Z|\mathcal{H}_1}(\zeta) \approx p_{Z|\mathcal{H}_0}(\zeta)$. In that case, the *a posteriori* SPP is not a function of the current observation anymore, as we obtain from Eq. (5.4)

$$P(\mathcal{H}_1 \mid \zeta)\Big|_{\xi_{\mathcal{H}_1}=0} = P(\mathcal{H}_1 \mid \zeta)\Big|_{p_{Z|\mathcal{H}_1}(\zeta)=p_{Z|\mathcal{H}_0}(\zeta)} = P(\mathcal{H}_1). \tag{5.10}$$

This is also illustrated in Figure 5.1 where we compare the likelihoods for different model parameters. Thus, for a short-term estimate of $\xi_{\mathcal{H}_1}$, e.g., obtained using the decision-directed approach, the case $p_{Z|\mathcal{H}_1}(\zeta) \ll p_{Z|\mathcal{H}_0}(\zeta)$ in Eq. (5.6) hardly occurs and usually $P(\mathcal{H}_1) \leq P(\mathcal{H}_1 \mid \zeta) \leq 1$.

5.2.2 FIXED OPTIMAL $\xi_{\mathcal{H}_1}$

In the radar or communication context the parameter $\xi_{\mathcal{H}_1}$ would be chosen to guarantee a specified performance in terms of false alarms or missed detections [22] in the environment it is used in. Also, in speech we have to deal with whatever SNR exists as a consequence of the particular acoustical

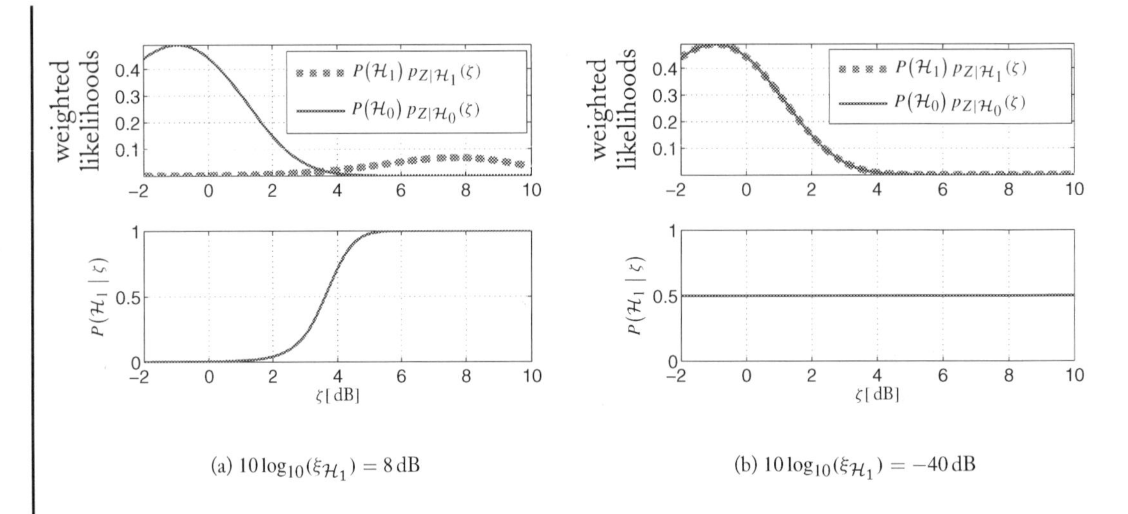

Figure 5.1: Likelihoods and *a posteriori* SPPs for large $\xi_{\mathcal{H}_1}$ (left) and $\xi_{\mathcal{H}_1} \rightarrow 0$ (right) for a smoothed observation. When the model parameter $\xi_{\mathcal{H}_1}$ is close to zero, it can be seen that the likelihoods overlap (right) such that the *a posteriori* SPP yields only the prior (here: $P(\mathcal{H}_1) = 0.5$) independent of the observation ζ.

environment. Thus, in [86] it is proposed to find an optimal, fixed $\xi_{\mathcal{H}_1}$ by minimizing false alarms and missed hits for a range of expected SNRs. Then, as opposed to a decision-directed estimate of $\xi_{\mathcal{H}_1}$, the likelihood function of speech presence does not reflect the underlying PDF of the current random variable Z in the time-frequency point under consideration, but rather the PDF that the random variable Z would have if speech were present in the considered time-frequency point. The resulting $\xi_{\mathcal{H}_1}$ can be interpreted as a long-term SNR rather than the short-term SNR. As a result, in speech absence the likelihood of speech absence will be much lower than the likelihood of speech presence, and we have $P(\mathcal{H}_1 \mid \zeta) = 0$ as stated in Eq. (5.6). Thus, without any further modifications, choosing a fixed or long-term $\xi_{\mathcal{H}_1}$ enables SPP estimates close to zero and therefore a proper detection of speech absence.

5.3 CHOOSING THE PRIOR PROBABILITIES

5.3.1 ADAPTIVE PRIOR PROBABILITIES

As argued in Sec. 5.2.1 and Eq. (5.10), the *a posteriori* probability of speech presence tends to $P(\mathcal{H}_1)$ rather than to zero in speech absence when $\xi_{\mathcal{H}_1}$ is an estimate of the current local SNR. To improve this undesired behavior, Malah et al. [85] proposed to estimate the *a priori* SPP as a function of the observation, e.g., by an iterative procedure. However, under speech absence the resulting *a posteriori* SPP remains close to 0.5 instead of taking values close to zero (see Figure 5.2(c)). Also Cohen and

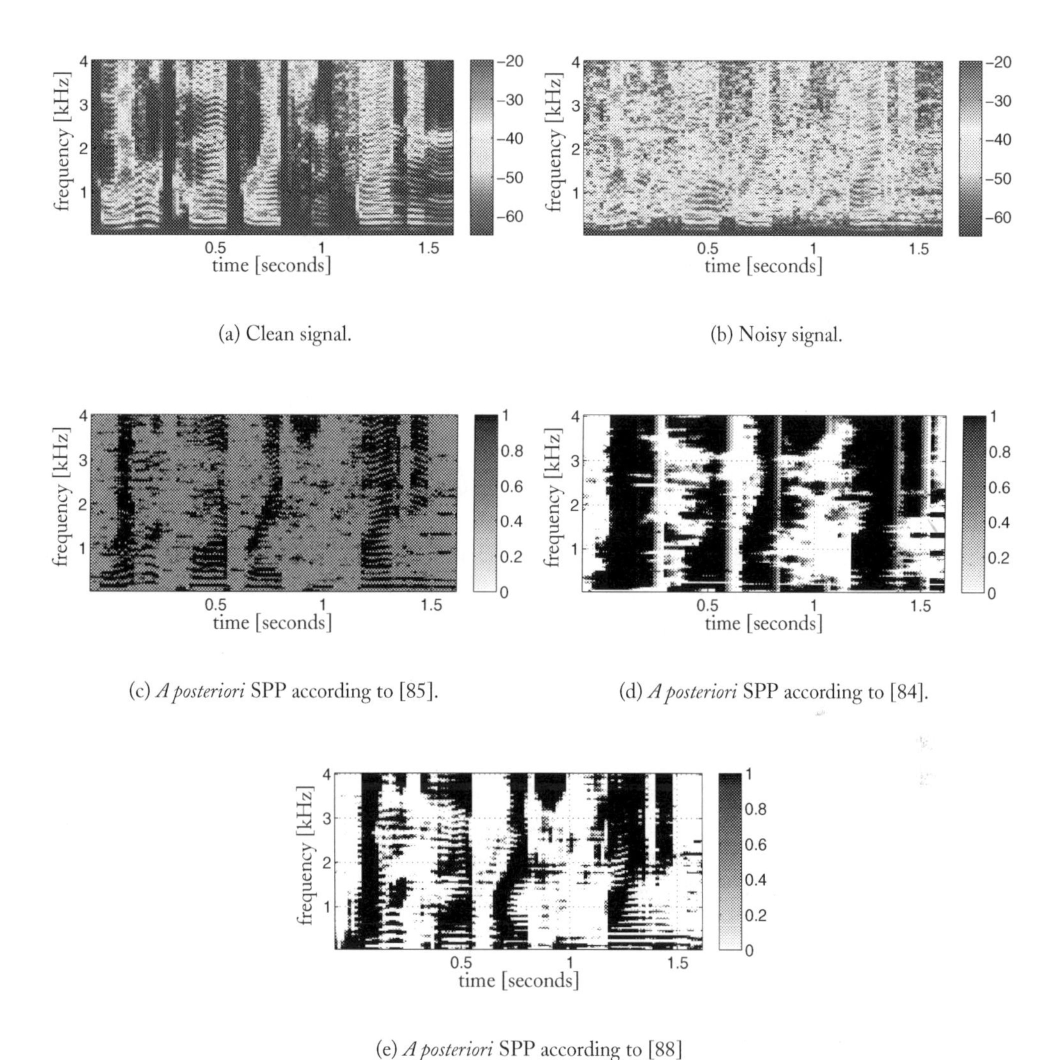

(a) Clean signal.

(b) Noisy signal.

(c) *A posteriori* SPP according to [85].

(d) *A posteriori* SPP according to [84].

(e) *A posteriori* SPP according to [88]

Figure 5.2: Example for the differences of SPP estimators. In (a) the spectrogram of clean speech is given while (b) shows noisy speech disturbed by babble noise. Figures (c)-(e) show the estimated SPP for different algorithms.

Berdugo [84] propose to adapt the priors. For this, they take averages of a decision-directed estimate of ξ, map these averages to values between 0 and 1, and reinterpret them as the *a priori* SPP. By doing so, they enable *a posteriori* SPP estimates close to zero also in speech absence even though $p_{Z|\mathcal{H}_1}(\zeta) = p_{Z|\mathcal{H}_0}(\zeta)$ and Eq. (5.10) holds instead of Eq. (5.6) (see Figure 5.2(d)).

5.3.2 FIXED PRIOR PROBABILITIES

As per their definition the *a priori* SPPs $P(\mathcal{H}_1) = 1 - P(\mathcal{H}_0)$ are not conditioned on the current observation, fixed values are often used for the prior probabilities. Furthermore, when a fixed or long-term SNR is used for $\xi_{\mathcal{H}_1}$, an adaptation of the prior is not necessary anymore [86]. The choice of $P(\mathcal{H}_1)$ biases the posterior probability either in the direction of speech presence or absence. Different values have been used in the literature. For instance, Ephraim and Malah use a large prior of $P(\mathcal{H}_1) = 0.8$ [14], while Kim and Chang use a value as low as $P(\mathcal{H}_1) = 0.0588$ [90]. A reasonable compromise is to assume that speech presence and absence are equally likely in each time-frequency point, i.e. $P(\mathcal{H}_1) = 0.5$. This assumption is used e.g. in [22] and the first iteration of [85], but also in the recent proposals [86, 88].

5.4 AVOIDING OUTLIERS

The methods proposed in [84, 86, 88] yield an estimate of the speech presence probability at each time-frequency point in the short-time frequency domain, and take values between zero and one. In a speech enhancement framework, outliers in the estimate can yield annoying artifacts. In [84] outliers are avoided by taking local and global averages of the *a priori* SNR. This idea is adopted in [86] where local and global averages of the *a posteriori* SNR are computed over time and frequency, as

$$\bar{\zeta}_{\Xi,k}(\ell) = \frac{1}{|\mathbb{K}_\Xi|\,|\mathbb{L}|} \sum_{\substack{\kappa \in \mathbb{K}_\Xi \\ \lambda \in \mathbb{L}}} \zeta_\kappa(\lambda)\,, \tag{5.11}$$

where Ξ stands for either 'local' or 'global'. While $\mathbb{L}$ is chosen to corresponds to the past 64 ms, $\mathbb{K}_{\mathrm{local}}$ corresponds to a frequency band of approximately 100 Hz, while $\mathbb{K}_{\mathrm{global}}$ corresponds to approximately 550 Hz. In [84, 86] the overall probability estimate is obtained by multiplying probabilities obtained by local and global averages, for instance [86]

$$\widehat{P}(\mathcal{H}_1 \mid \zeta) = P(\mathcal{H}_1 \mid \zeta_{\mathrm{local}})\, P(\mathcal{H}_1 \mid \zeta_{\mathrm{global}})\,. \tag{5.12}$$

As in the cepstral domain a selective smoothing of spectral outliers is facilitated, in [88] the performance is increased by smoothing the *a posteriori* SNR in the cepstral domain similar to Sec. 7.3. Using a temporal cepstrum smoothing also makes the heuristic combination of local and global averages in Eq. (5.12) unnecessary [88]. In both [86, 88] the PDFs of the *a posteriori* SNR after smoothing can be well modeled by an increase in the shape parameter ν in Eq. (5.7) and Eq. (5.8) [86, 91]. In Figure 5.2(e) it can be seen that in contrast to [85] the cepstral approach [88] yields values close to zero in speech absence and better resolves the spectral harmonics of voiced speech than [84].

CHAPTER 6

Noise PSD Estimation

From Sec. 4 it follows that speech DFT-coefficient estimators are generally a function of the noise PSD σ_N^2, either directly, or, indirectly via the *a priori* and *a posteriori* SNR. The quality of the estimated speech signal therefore heavily depends on the accuracy of the noise PSD estimate. The noise PSD can be underestimated or overestimated. An underestimate of σ_N^2 generally leads to an undersuppression of the noisy speech, and an unnecessarily large amount of residual noise. On the other hand, an overestimate of σ_N^2 generally leads to an oversuppression of the noisy speech, and a potential loss in speech quality or intelligibility. Noise PSD estimation is thus a crucial aspect of any noise reduction algorithm for speech enhancement, and is in particular challenging when speech is degraded with non-stationary noise.

Before going into the details of noise PSD estimation, it is useful to recall that σ_N^2 is defined as the expected value of the random variable $|N|^2$, that is, $\sigma_N^2 = E(|N|^2)$. In practice, however, we never have access to this expected value, even if we would have access to the noise realization. In fact, the situation is even worse, as we can generally only observe a single realization of the *noisy* process, not the noise process. Despite these apparent difficulties, quite a lot of methods exist for estimating the noise PSD σ_N^2 from a noisy signal realization.

To demonstrate the strengths and weaknesses of the different noise PSD estimators we degrade a speech signal (duration 35 s) originating from the Timit database [67] with additive speech-shaped noise at a global SNR of 10 dB. In order to verify how well the different noise PSD tracking approaches are able to handle stationary as well as less stationary noise types, the speech shaped noise is modulated with an increasing modulation frequency from 0 Hz to 0.5 Hz for 25 seconds. Then, after 25 seconds the noise level stays constant for about two seconds, after which two sinusoids with frequencies of 600 and 1200 Hz are added to verify the noise tracking performance when deterministic noise types are also present. In Figure 6.1 the true noise PSD of this noise source is shown for the frequency bin that is representative for the sinusoid at 1200 Hz, together with the estimated noise PSDs for the methods that will be discussed.

6.1 METHODS BASED ON VAD

Early methods for noise PSD estimation exploited the fact that between words, and even between syllables, speech is absent for a short moment. During these moments, the noisy signal degenerates to the noise realization itself, enabling estimation of the noise PSD. To detect these non-speech parts many methods have been proposed, which are often referred to as voice activity detectors (VADs). Some of these methods, e.g., [92], rely on the fact that certain statistics, e.g., the energy or the log-energy of the noise-only process and the noisy speech process, are different. By comparing these statistics or their descriptors like expectation and variance to the actual energy or log-energy of the signal, it can be decided whether speech is absent or present. As in Sec. 5, let the two hypotheses $\mathcal{H}_1(\ell)$ and $\mathcal{H}_0(\ell)$ indicate that speech is present and absent, respectively, in a particular time-frame ℓ. The noise PSD estimate $\widehat{\sigma}_{\mathrm{N}}^2$ can then be obtained by recursive smoothing, i.e.,

$$\widehat{\sigma}_{\mathrm{N},k}^2(\ell) = \begin{cases} \alpha\widehat{\sigma}_{\mathrm{N},k}^2(\ell-1) + (1-\alpha)|y_k(\ell)|^2 & \text{when } \mathcal{H}_1(\ell) \\ \widehat{\sigma}_{\mathrm{N},k}^2(\ell-1) & \text{when } \mathcal{H}_0(\ell), \end{cases} \tag{6.1}$$

where $0 \leq \alpha \leq 1$ is a smoothing constant. Although VADs are conceptually relatively simple, their ability to accurately estimate the noise PSD and track changes in non-stationary noise is rather limited. More specifically, when the noise PSD changes while speech is present, $\widehat{\sigma}_{\mathrm{N},k}^2(\ell)$ will not be updated. As this is not uncommon to happen, e.g., consider the case of a passing car, there is a need to also update $\widehat{\sigma}_{\mathrm{N},k}^2(\ell)$ while speech is present in a particular time-frame.

In Figure 6.1(a) we demonstrate the noise PSD estimator based on the VAD presented in [92]. This VAD determines whether speech is present or absent by comparing the (estimated) log-energy of the noise per time-frame to dynamically adjusted thresholds. As is clear from Figure 6.1(a), this VAD-based approach is not able to follow the quick changes in the noise level. The reason for this is that the noise level changes too quickly (even for a low modulation frequency) and during speech presence.

6.2 METHODS BASED ON MINIMUM POWER LEVEL TRACKING

Many noise PSD estimators exploit the fact that even when speech is present in a certain time-frame, speech energy is not necessarily present in each and every frequency bin of that time-frame. Voiced speech sounds, for example, have a quasi-periodic time-domain signal and thus a quasi-harmonic power spectrum, with spectral peaks located at integer multiples of the inverse of the period. The spectral content between these spectral peaks is representative for the noise level. In Figure 6.2(a) this is demonstrated in the spectrogram of a noisy speech signal consisting of clean speech degraded with white Gaussian noise at 20 dB SNR. In this spectrogram we can easily distinguish the different speech sounds, and, observe that for the majority of the speech sounds, speech energy is mainly present at regularly spaced intervals at lower frequencies.

Figure 6.2(b) shows a cross-section of the spectrogram from Figure 6.2(a) at a frequency of approximately 520 Hz. In addition, the true value of the noise PSD at that particular frequency bin is shown. This cross-section shows that when a speech-related spectral peak is present in the frequency bin, the noisy speech power rises far above the noise level. However, in many time-frames, the energy level varies around the true noise level.

The minimum statistics (MS) method by Martin [38, 93] uses this observation to estimate the noise PSD without using an explicit VAD. The idea of this approach is to collect smoothed

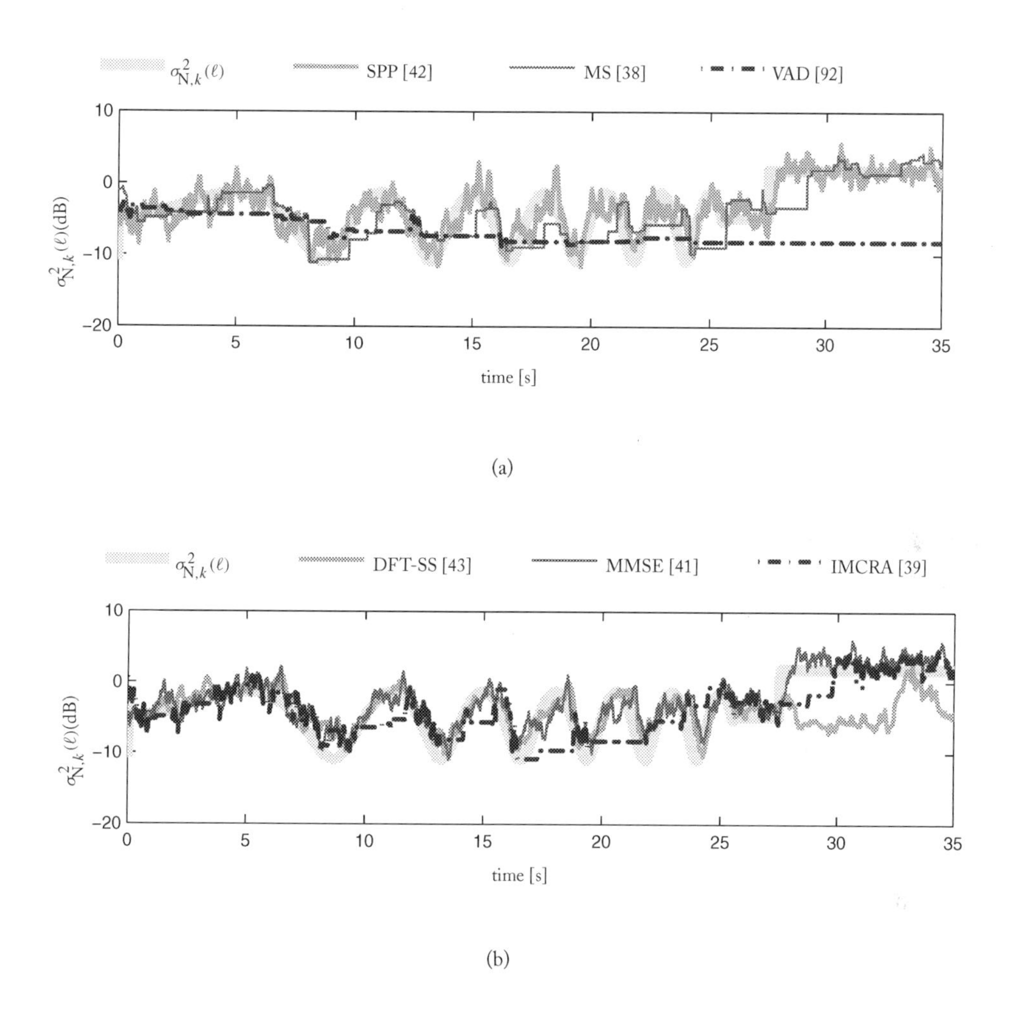

Figure 6.1: Estimated noise PSD and true noise PSD for frequency bin centered at approximately 1203 Hz.

noisy periodogram values, $Q_k(\ell)$, i.e.,

$$Q_k(\ell) = \alpha_k(\ell) Q_k(\ell - 1) + (1 - \alpha_k(\ell)) |y_k(\ell)|^2, \tag{6.2}$$

for a particular frequency bin in a sequence of L time-frames, that is,

$$\mathbf{Q} = \{Q_k(\ell - L + 1), \ldots, Q_k(\ell)\}. \tag{6.3}$$

By making L sufficiently large, i.e., corresponding to around 1 to 2 seconds, it is guaranteed that the minimum value in the sequence $\mathbf{Q}$, say $Q_{\min}$, originates from a time-frequency point without

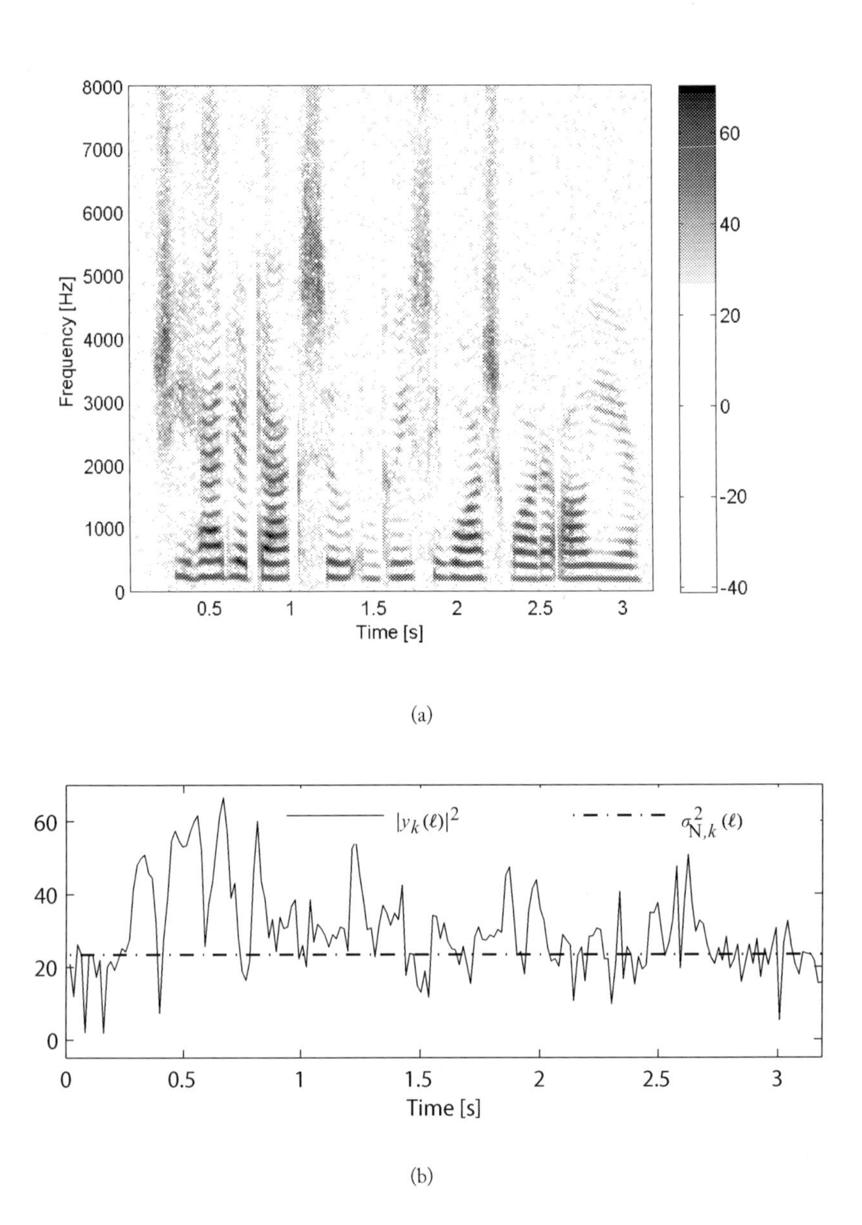

Figure 6.2: (a) Spectrogram of a Timit database sentence degraded with additive white Gaussian noise at 20 dB SNR. (b) Noisy speech power in frequency bin centered around 520 Hz.

speech presence. Relating this to Figure 6.2(b), $\mathbf{Q}$ could represent a smoothed version of the values during an interval of 1.5 seconds. However, by taking the minimum of $\mathbf{Q}$, in general the distribution of $|N|^2$ is sampled from below its mean, i.e., $Q_{\min}$ is an underestimate of $\mathrm{E}(|N|^2)$. In order to correct for this, a bias compensation is necessary, leading to $\widehat{\sigma}_{\mathrm{N}}^2 = B_{\min} Q_{\min}$, where $B_{\min}^{-1} = \mathrm{E}(Q_{\min}|\sigma_{\mathrm{N}}^2 = 1)$ is the bias compensation for which approximate solutions are given in [38]. Although the concept of the MS approach is relatively simple, a successful implementation requires quite a number of additional parameter settings. From Figure 6.1(a) it can be seen that when the noise level is slowly changing, e.g., during the period from 0 to 7 seconds, MS is quite accurate. Also when the (stationary) sinusoidal noise components are present, quite an accurate noise PSD estimate is obtained. However, a drawback of the MS approach is that by computing the spectral minimum of $L-1$ past frames, detection of an increasing noise level at a frequency bin has a worst case delay of $L-1$ frames. Depending on parameter settings, this delay could be as large as one to two seconds. This is clearly visible in Figure 6.1(a) when the noise modulations become faster and when the sinusoidal component is added after 27 seconds. Such fast and abrupt changes in the noise can generally not be tracked accurately with the MS approach, resulting in a large amount of residual noise due to an underestimated noise PSD.

6.3 SPP-BASED NOISE PSD ESTIMATION

Another class of noise PSD estimation algorithms exploits the *a posteriori* SPP discussed in Sec. 5, e.g., [39, 40, 42, 89]. Using the *a posteriori* SPP $P(\mathcal{H}_1 \mid Y_k(\ell))$, an MMSE estimator of the noise PSD is given by

$$\begin{aligned}\mathrm{E}\left(\sigma_{\mathrm{N},k}^2(\ell)|Y_k(\ell)\right) =& \mathrm{E}\left(\sigma_{\mathrm{N},k}^2(\ell)|Y_k(\ell), \mathcal{H}_1\right) P(\mathcal{H}_1 \mid Y_k(\ell)) \\ &+ \mathrm{E}\left(\sigma_{\mathrm{N},k}^2(\ell)|Y_k(\ell), \mathcal{H}_0\right) \left(1 - P(\mathcal{H}_1 \mid Y_k(\ell))\right).\end{aligned} \tag{6.4}$$

In [89], it was proposed to approximate the conditional expectations $\mathrm{E}\left(\sigma_{\mathrm{N},k}^2(\ell)|Y, \mathcal{H}_1\right)$ and $\mathrm{E}\left(\sigma_{\mathrm{N},k}^2|Y, \mathcal{H}_0\right)$ in Eq. (6.4) as $\widehat{\sigma}_{\mathrm{N},k}^2(\ell-1)$ and $|y_k(\ell)|^2$, respectively. In addition, it was proposed to compute one SPP per time-frame instead of one for each frequency bin. To do so, instead of computing $p_{Z_k(\ell)|\mathcal{H}_0}(\zeta_k(\ell))$ and $p_{Z_k(\ell)|\mathcal{H}_1}(\zeta_k(\ell))$ in Eq. (5.4) per frequency bin, it was assumed that the realizations ζ of the *a posteriori* SNR Z are statistically independent across frequency after which the likelihoods are combined using the geometric mean, that is, $\overline{p_{Z_k(\ell)|\mathcal{H}_0}(\zeta_k(\ell))} = \sqrt[K]{\prod_{\forall k} p_{Z_k(\ell)|\mathcal{H}_0}(\zeta_k(\ell))}$ and $\overline{p_{Z_k(\ell)|\mathcal{H}_1}(\zeta_k(\ell))} = \sqrt[K]{\prod_{\forall k} p_{Z_k(\ell)|\mathcal{H}_1}(\zeta_k(\ell))}$, respectively, where $\bar{\cdot}$ indicates the geometric mean. Substituting these likelihoods into Eq. (5.4), the frequency independent *a posteriori* SPP $\overline{P(\mathcal{H}_1 \mid \zeta)}$ is obtained. Altogether, it was proposed in [89] to estimate the noise PSD by means of

$$\sigma_{\mathrm{N},k}^2(\ell) = \widehat{\sigma}_{\mathrm{N},k}^2(\ell-1)\overline{P(\mathcal{H}_1 \mid \zeta)} + |y_k(\ell)|^2 \left(1 - \overline{P(\mathcal{H}_1 \mid \zeta)}\right). \tag{6.5}$$

Notice that in contrast to Eq. (6.1) this is a soft decision between speech presence and speech absence, although the speech presence probability is still frequency independent. This method is not very robust, as it might happen quite regularly that a time-frame is estimated to be noise-only while speech is actually present, leading to situations where speech leaks into the noise PSD estimate.

A further improvement of SPP-based noise PSD estimation was proposed by Cohen in [39], who estimated the noise PSD by combining the notion of SPP with recursive averaging. In contrast to [89], the SPPs were made time-frequency dependent and the conditional expectation under speech absence $E\left(\sigma_{N,k}^2 | Y, \mathcal{H}_0\right)$ was approximated as $\alpha\widehat{\sigma}_{N,k}^2(\ell-1) + (1-\alpha)|y_k(\ell)|^2$, so that the MMSE estimate of the noise PSD becomes

$$\begin{aligned}\widehat{\sigma}_{N,k}^2(\ell) =& P\left(\mathcal{H}_1 \mid Y_k(\ell)\right) \widehat{\sigma}_{N,k}^2(\ell-1) + \left(1 - P\left(\mathcal{H}_1 \mid Y_k(\ell)\right)\right) \\ & \left(\alpha\widehat{\sigma}_{N,k}^2(\ell-1) + (1-\alpha)|y_k(\ell)|^2\right).\end{aligned} \tag{6.6}$$

Setting $\tilde{\alpha}_k(\ell) = \alpha + (1-\alpha)P\left(\mathcal{H}_1 \mid Y_k(\ell)\right)$, Eq. (6.6) can be reformulated as [39]

$$\widehat{\sigma}_{N,k}^2(\ell) = \tilde{\alpha}_k(\ell)\widehat{\sigma}_{N,k}^2(\ell-1) + (1-\tilde{\alpha}_k(\ell))|y_k(\ell)|^2, \tag{6.7}$$

which shows that the smoothing parameter $\tilde{\alpha}_k(\ell)$ is in fact time-frequency dependent and completely determined by the speech presence probability $P\left(\mathcal{H}_1 \mid Y_k(\ell)\right)$. This probability is defined as in Eq. (5.9) with $\nu = 1$, while the *a priori* probability of speech absence $P(\mathcal{H}_0)$ in Eq. (5.9) is made time-frequency dependent and adaptive with respect to the observation. However, notice that *a priori* probabilities are per definition not conditioned on the observation (see also Sec. 5.3.2). Nevertheless, the exact value of the adaptive $P(\mathcal{H}_0)$ as proposed in [39] depends on a minimum tracking procedure similar to the MS approach [38], explaining the name improved minima controlled recursive averaging (IMCRA) of this method. The noise PSD estimated using IMCRA is shown in Figure 6.1(b). In particular, during the first 7 seconds and during the last 5 seconds, the noise PSD estimates are relatively accurate. However, similarly to the MS approach, IMCRA has difficulties following quickly increasing noise levels due to the fact that it is also based on the same minimum statistics principles.

In [42] an improved SPP-based noise PSD estimation was proposed. This method can also be formulated using Eq. (6.4) and made the same approximations for the conditional expectations. However, in contrast to [89] and [39], it was proposed to employ an improved *a posteriori* SPP based on a fixed non-adaptive *a priori* SNR $\xi_{\mathcal{H}_1}$ as explained in Sec. 5.2.2. Using this fixed *a priori* SNR, close to zero *a posteriori* SPP values $P\left(\mathcal{H}_1 \mid Y_k(\ell)\right)$ can be obtained in the case of speech absence, allowing a correct adaptation of the noise PSD. Further, choosing fixed priors $\xi_{\mathcal{H}_1}$ and $P(\mathcal{H}_1)$ has the advantage that the noise PSD estimation is decoupled from any subsequent steps of the speech enhancement framework, such as the estimation of the *a priori* SNR or the estimation of clean speech spectral coefficients.

In Figure 6.1(a) it is shown that this SPP-based estimator indeed leads to rather accurate noise PSD estimates in regions where the noise source is stationary, as well as in regions where the noise source is non-stationary. At the same time, this approach exhibits a very low computational complexity and memory usage, making it particularly suitable for mobile devices.

6.4 MMSE-BASED ESTIMATION OF THE NOISE PSD

Another group of noise PSD estimators approaches the problem from the viewpoint of Bayesian MMSE estimation, e.g., [41, 94]. These approaches estimate the noise PSD by a recursive smoothing of an MMSE estimate of the noise periodogram, that is the conditional expectation, $\mathrm{E}\left(|N_k(\ell)|^2 | y_k(\ell)\right)$, which can be computed based on assumed distributions of speech and noise DFT coefficients. The resulting estimator is a function of both the *a priori* and *a posteriori* SNR, say $\mathrm{E}\left(|N_k(\ell)|^2 | y_k(\ell)\right) = f\,(\xi_k(\ell), \zeta_k(\ell))$. Since both $\xi_k(\ell)$ and $\zeta_k(\ell)$ are functions of the noise PSD itself, it is common to replace this with the noise PSD estimate of the previous frame, i.e., $\widehat{\sigma}^2_{\mathrm{N},k}(\ell - 1)$. The main difference between the approaches presented in [94] and [41] is the way in which $\xi_k(\ell)$ is estimated in order to compute $f\,(\xi_k(\ell), \zeta_k(\ell))$. In both [94] and [41] it was observed that an estimated $\xi_k(\ell)$ leads to a biased noise PSD estimate. In [41] this is solved by computing an analytic expression for the bias, while in [94] this was solved based on a more heuristically motivated procedure.

The MMSE approach presented in [41] has a rather good tracking performance for both stationary and non-stationary noise sources. From Figure 6.1(b) we see that its performance is similar to the SPP approach in Figure 6.1(a), while its complexity is slightly higher compared to the SPP approach [42].

6.5 DFT-SUBSPACE ESTIMATION OF THE NOISE PSD

The final algorithm that will be discussed in this section is the DFT-subspace approach presented in [43, 95]. This method exploits the tonal structure of voiced speech sounds. Where other methods, like MS [38], only employ the noise-only frequency bins e.g., between harmonics, this method is also able to estimate the noise PSD for frequency bins where speech energy is constantly present, e.g., in frequency regions of a harmonic. The assumption that underlies this method is that a speech signal observed in a particular frequency bin across time is confined to a low-dimensional vector subspace within the higher-dimensional vector space of the noisy signal. More specifically, by constructing correlation matrices of time-series of complex-DFT coefficients per frequency bin, these can be decomposed using an eigenvalue decomposition into two matrices of which the columns span two orthogonal vector spaces. One of these two subspaces is a noise-only subspace. The noise-only eigenvalues of the corresponding matrix can be used to update the noise PSD. This method has been shown to be quite powerful in estimating the noise PSD of fast varying noise sources [96]. This is visible in Figure 6.1(b), where a rather accurate tracking of the true noise level is shown. However, due to the eigenvalue decompositions per time-frequency point, the computational complexity is higher than for other methods described in this section, e.g., compared to the MMSE approach presented in [41] complexity is about a factor 40 higher [41]. Another limitation of this approach is that due to the assumed low-rank speech model, noise types that obey a low-rank model themselves, e.g., periodic noises such as the harmonics of an engine, will not be present in the noise subspace and have to be estimated using additional measures. This is clearly visible in Figure 6.1(b). When the sinusoidal noise components are added after approximately 27 seconds, the estimated noise level stays around the same level as the stochastic part of the noise, which is stationary from 25 till 35 seconds.

CHAPTER 7

Speech PSD Estimation

In this section we review state-of-the art speech PSD estimators, block C) in Figure 2.1, and discuss their strengths and weaknesses.

Similar to the noise PSD, the speech PSD is defined as $\sigma_S^2 = \mathrm{E}(|S|^2)$. Also for the speech PSD, we have the problems that only the noisy signal $Y = S + N$ can be observed and that we only observe one realization of the random variable in each time frequency point such that computing the expected value is problematic. As speech is typically even more non-stationary than noise, we cannot simply replace the expected value operator by a recursive smoothing. On the other hand, using the magnitude square of the spectral coefficients, often referred to as the *periodogram*, as an estimate of the PSD exhibits a large variance. Thus, in order to reduce the variance of the speech PSD estimate, careful, possibly adaptive smoothing is applied that preserves the properties of speech as well as possible.

7.1 MAXIMUM LIKELIHOOD ESTIMATION AND DECISION-DIRECTED APPROACH

A simple estimate of the speech PSD can be found by ML estimation. For this, we need to maximize the likelihood function $p_{Y|\sigma_S^2}(y \mid \sigma_S^2)$ with respect to the speech PSD. Assuming that the speech and noise spectral coefficients are complex-Gaussian distributed, the optimal ML-estimator is found to be [14]

$$\sigma_{S,k}^{2,\mathrm{ML}}(\ell) = |y_k(\ell)|^2 - \sigma_{N,k}^2(\ell). \tag{7.1}$$

If speech and noise are uncorrelated, this is an unbiased estimate of the speech spectral power, as $\mathrm{E}\left(\sigma_S^{2,\mathrm{ML}}\right) = \mathrm{E}(|S + N|^2) - \sigma_N^2 = \sigma_S^2$. Because the ML estimate of the noisy periodogram in Eq. (7.1) does not contain any smoothing, it exhibits a large variance. This is not desired, as random fluctuations in $|y|^2$ cause isolated spectral peaks in the estimated speech spectrum when the ML estimate Eq. (7.1) is used as an estimate of the speech PSD in a speech enhancement framework. These isolated spectral peaks correspond to local sinusoids in the time domain and are perceived as tonal artifacts of one frame duration, often referred to as *musical noise*. The effect of musical noise can be reduced by considering a sequence of successive periodograms. Assuming successive periodogram bins are independent and have a common speech and noise PSD, the ML-estimate for a sequence of length L is given by [14]

$$\sigma_{S,k}^{2,\mathrm{ML}}(\ell) = \frac{1}{L}\sum_{n=0}^{L-1} |y_k(\ell - n)|^2 - \sigma_{N,k}^2(\ell). \tag{7.2}$$

Thus, under the given assumptions the optimal ML solution is obtained by taking the average over successive noisy periodograms, and subtracting the noise PSD from the average.

However, since speech is non-stationary the assumption that L successive segments have the same speech PSD is often violated in practice. Thus, the averaging of the noisy periodogram results in a poor trade-off between fluctuations in the residual noise and distortions of speech onsets and transients. If L is chosen large enough to eliminate fluctuations in $\sigma_S^{2,\mathrm{ML}}$, it also distorts speech onsets and transitions, resulting in reduced speech quality. To obtain a better trade-off, Ephraim and Malah proposed a *decision-directed* (DD) estimator based on a previous clean-speech estimate $\widehat{S}_k(\ell - 1)$, which can be implemented as [14, 34]

$$\widehat{\sigma_{S,k}^2}(\ell) = \max\left(\alpha_{\mathrm{dd}}|\widehat{S}_k(\ell-1)|^2 + (1-\alpha_{\mathrm{dd}})\left(|y_k(\ell)|^2 - \sigma_{N,k}^2(\ell)\right), \xi_{\min}\sigma_{N,k}^2(\ell)\right). \tag{7.3}$$

The parameters α_{dd} and $\xi_{\min}$ control the trade-off between noise reduction and distortions of speech transients in a speech enhancement framework [97]. The decision-directed procedure Eq. (7.3) allows for fast tracking of increasing levels of the speech power, and effectively results in an adaptive smoothing. Consequently, at speech onsets and transitions, fewer speech distortions are introduced as compared to the averaging in Eq. (7.2), while still yielding fewer outliers than the instantaneous estimate in Eq. (7.1). The decision-directed approach is probably the most widely used speech PSD estimator. Its performance has been discussed by various authors (see for instance [16, 48, 97] for rigorous analyses).

However, the problem of musical noise is not completely overcome when the decision-directed approach is used: since it is sensitive to rising spectral amplitudes, it does not only respond to speech onsets, but also to outliers in the noise signal. Therefore, outliers in the noise signal are likely to cause outliers in the residual noise of the clean-speech estimate that may be perceived as annoying musical tones. Furthermore, the smoothing properties of the decision-directed approach depend on the type of speech estimator employed [48].

7.2 KALMAN-TYPE FILTERING, GARCH MODELING, AND NONCAUSAL ESTIMATION

While initially the decision-directed approach was heuristically motivated, Cohen [16] showed that it can be interpreted as a special case of a Kalman-type filter based estimation of the speech PSD with an adaptive smoothing parameter. For this, Cohen decomposed the estimation into a propagation step and an update step. In [16] the propagation step is heuristically chosen as a weighted averaging of a conditional power estimate $E(A^2|Y)$ and the square of an estimate of the clean speech spectral magnitude.

- Propagation:

$$\widetilde{\sigma_S^2}(\ell) = \max\left\{(1-\alpha_{\mathrm{pr}})\,\widehat{\sigma_S^2}(\ell-1) + \alpha_{\mathrm{pr}}\widehat{A}^2(\ell-1),\, \xi_{\min}\sigma_N^2\right\} \tag{7.4}$$

- Update:

$$\widehat{\sigma_S^2}(\ell) = \mathrm{E}\Big(A^2(\ell) \mid Y(\ell), \widetilde{\sigma_S^2}(\ell)\Big) \tag{7.5}$$

- Application:

$$\widehat{A}(\ell) = c^{-1}\Big(\mathrm{E}\Big(c(A) \mid Y(\ell), \widehat{\sigma_S^2}(\ell)\Big)\Big), \tag{7.6}$$

where a typical choice for the function $c(\cdot)$ is the logarithm $c(\cdot) = \log(\cdot)$ which results in the log-spectral amplitude estimator [37]. An estimator for the conditional power in Eq. (7.5) is found e.g. in [72]. Cohen shows that for $\alpha_{\mathrm{pr}} = 1$ the Kalman-type estimator results in a decision-directed estimator with a frequency dependent smoothing factor, and proposes to choose $\alpha_{\mathrm{pr}} = 0.9$.

In [34] Cohen replaces the heuristic propagation step in Eq. (7.4) by a so-called generalized autoregressive conditional heteroscedasticity (GARCH) model resulting in the propagation step

$$\widetilde{\sigma_S^2}(\ell) = \sigma^2_{S,\min} + \alpha_{\mathrm{GARCH}}\widehat{\sigma_S^2}(\ell-1) + \beta_{\mathrm{GARCH}}\left(\widetilde{\sigma_S^2}(\ell-1) - \sigma^2_{S,\min}\right). \tag{7.7}$$

A drawback of the GARCH model proposed in [34] is that the parameters α_{GARCH}, β_{GARCH}, and $\sigma^2_{S,\min}$ are trained offline for each individual speaker, which reduces the flexibility of the algorithm.

If L future segments are available, speech PSD estimation can be improved by formulating a noncausal estimator, e.g. [16]. However, the additional delay caused by the look-ahead of the noncausal estimator is not tolerable in many real-time communication applications such as hearing aids.

7.3 TEMPORAL CEPSTRUM SMOOTHING

As we have argued above, estimating the speech PSD is usually a trade-off between removing undesired fluctuations in the estimate and an undesired smearing of the speech signal. In general, isolated spectral peaks in the ML estimate $\sigma^{2,\mathrm{ML}}_{S,k}$ that would cause musical noise in a speech enhancement framework are not likely to be produced by humans. Thus, a better estimate of the speech PSD can be expected, if more *a priori* knowledge about the spectral properties of speech signals is employed. To do so, in [98] it is proposed to smooth the ML estimate $\sigma^{2,\mathrm{ML}}_{S,k}$ in the cepstral domain. The cepstrum is given by an inverse Fourier transform of the logarithmically compressed speech PSD estimate across frequency. Then, the lower cepstral coefficients represent the spectral envelope of the compressed speech PSD estimate, while the upper coefficients represent its spectral fine structure. To have a good representation of speech signals, we need to preserve the speech spectral envelope—which is of predominant importance to distinguish between speech sounds—and the speech spectral harmonics in voiced sounds. In the cepstrum, the speech spectral envelope is represented by only few lower cepstral coefficients, while the regular structure of the speech spectral harmonics is mapped mainly to a single peak in the cepstrum. An example of a sequence of short-term speech periodograms and the corresponding cepstra can be seen in Figure 7.1(a) and Figure 7.1(b). On the other hand, spectral outliers due to estimations errors are likely to be represented by a different set of cepstral coefficients. Thus, in the cepstral domain a selective smoothing of speech related and non-speech related cepstral coefficients is possible. The steps of temporal cepstrum smoothing (TCS)-based speech PSD estimation are as follows:

- obtain the ML estimate $\sigma_{S,k}^{2,ML}(\ell)$ via Eq. (7.1),
- compute the cepstral representation

$$\lambda_q^{ML} = \mathrm{IDFT}_q\left\{ \log\left(\max\left(\sigma_{S,k}^{2,ML}, \sigma_N^2 \xi_{\min}\right)\right)\right\},$$

- find the fundamental period q_0, by searching for the maximum in $\lambda_q^{ML}(\ell)$ for $q > f_s/300$ Hz, where f_s is the sampling rate,
- selective temporal smoothing

$$\overline{\lambda}_q(\ell) = \alpha_q(\ell)\overline{\lambda}_q(\ell-1) + (1-\alpha_q(\ell))\lambda_q^{ML}(\ell),$$

  where α_q is close to one for the upper cepstral coefficients except q_0, and closer to zero for the remaining cepstral coefficients,
- transform back to frequency domain to obtain the speech PSD estimate

$$\widehat{\sigma_{S,k}^2} = \mathcal{B}\exp\left(\mathrm{DFT}_k\{\overline{\lambda}_q\}\right),$$

  where the factor $\mathcal{B}$ compensates for the bias that occurs because of the averaging in the log-domain. While in [98] a fixed value for $\mathcal{B}$ has been applied, in [91] an improved bias compensation is proposed which is a function of the smoothing factor α_q.

7.4 COMPARISON OF THE ESTIMATORS

In Figure 7.2 we compare the results of different speech PSD estimators for babble noise at 5 dB global input SNR. In order to draw conclusions independently of the chosen noise power estimator, for this experiment we use an ideal noise PSD, obtained by smoothing the noise periodogram over time. Figure 7.2(a) shows the periodogram of the noisy input, while Figure 7.1(a) shows the corresponding periodogram of the clean speech signal, which is unknown in practice. It can be seen that the decision-directed approach in Figure 7.2(b) yields localized undesired peaks of high energy where speech is absent, e.g., in the first 0.9 seconds. These isolated overestimations of the speech PSD are likely to result in musical noise in a speech enhancement framework. For the GARCH and

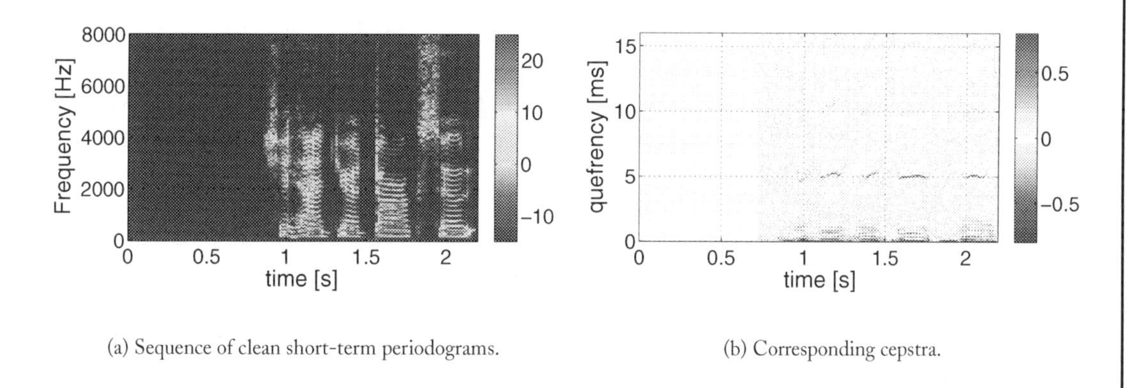

(a) Sequence of clean short-term periodograms.

(b) Corresponding cepstra.

Figure 7.1: Sequence of short-term speech periodograms and corresponding cepstra for a female speaker with $f_0 \approx 200\,\text{Hz}$.

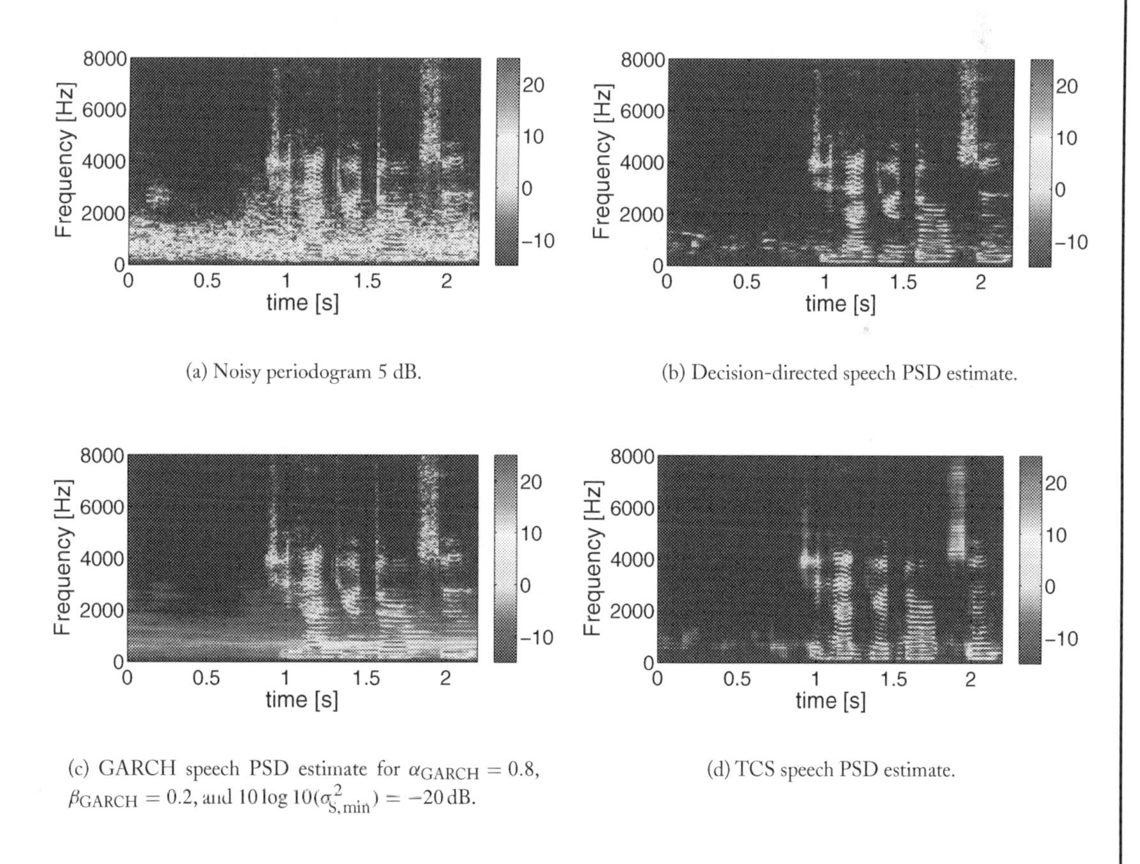

(a) Noisy periodogram 5 dB.

(b) Decision-directed speech PSD estimate.

(c) GARCH speech PSD estimate for $\alpha_{\text{GARCH}} = 0.8$, $\beta_{\text{GARCH}} = 0.2$, and $10 \log 10(\sigma^2_{S,\min}) = -20\,\text{dB}$.

(d) TCS speech PSD estimate.

Figure 7.2: Speech PSD estimates for babble noise a 5 dB global input SNR.

the choice of $\alpha_{\text{GARCH}} = 0.8$, $\beta_{\text{GARCH}} = 0.2$, we do not have isolated overestimations, however, the speech PSD tends to be overestimated in noise-only regions which will result in less noise reduction (Figure 7.2(c)). Also for the TCS approach some noise energy leaks into the speech PSD estimate. However, isolated spectral peaks are well reduced, meaning that the estimator is likely to result in less musical noise. At the same time, the speech spectral structure is well preserved (Figure 7.2(d)).

CHAPTER 8

Performance Evaluation Methods

Evaluating the performance of speech enhancement systems is important both during system development, to gauge the progress of the development work, and to allow comparison of performance across systems. For speech enhancement systems intended for human end-users (as opposed to machine-recognizers), it is common to ask a group of typical listeners to judge the resulting signals, i.e., to conduct a listening test. More specifically, the listening test subjects are often asked to focus on two attributes associated with speech signals in general, namely, quality and intelligibility.

The quality aspect of speech is hard and time consuming to evaluate accurately. The difficulty arises because speech quality may be characterized using several different dimensions or attributes such as "naturalness," "pleasantness," etc., because individual listeners may place different emphasis on each such dimension, and because quality evaluation procedures often measure the combined perception of *several* dimensions. Nevertheless, listening test procedures for evaluating quality aspects have been standardized to improve reproducibility and ease comparison of test results.

Intelligibility, on the other hand, is easier to define: how much (i.e., how many words, phonemes, etc.) of a given speech utterance was understood? It is also easier to measure reliably in listening experiments, because there is no subjective preference involved.

Listening tests may be considered the only tool for revealing the "true" performance of an enhancement system, but they are time consuming and costly to conduct. In the pursuit of cheaper and simpler alternatives, several *objective* or *instrumental distortion measures* have been developed, which compare the noisy or enhanced speech signal to a noise free reference signal, and produce an output, often a single scalar, which, ideally, correlates well with listening test results. Although not a substitute for listening tests, useful instrumental distortion measures exist, which can serve as a guiding tool during system development.

In the following we briefly touch upon subjective and instrumental methods for evaluating speech enhancement systems. An exhaustive coverage is far from possible, so we simply outline methods, which we have found useful. For more comprehensive reviews, we refer to [99], [100, Chap. 9], for speech signal evaluation in general, and to [101, Chap. 10] which focuses more on speech signals generated by enhancement systems.

8.1 EVALUATING QUALITY ASPECTS OF ENHANCED SPEECH

8.1.1 LISTENING TESTS

Many standardized listening test procedures exist for evaluating speech quality in general (see e.g., [101] for a description of some of them). For example, in mean opinion score (MOS) tests, e.g., [102], listeners rate the *overall signal quality* on a numerical scale ranging from 1 to 5, where 5 indicates "excellent" and 1 indicates "bad" quality, see Table 8.1(c) for the full scale. The Diagnostic Acceptability Measure (DAM) tests [103] go one step further and ask the subjects to judge up to 16 attributes of the signal under test; these attributes describe the speech signal, the background sound, and the total quality, rather than just the overall quality asked for in MOS tests. Most of these standards were originally developed for evaluation of speech codec performance, but have nevertheless been applied to evaluation of speech enhancement systems.

The ITU-T P.835 standard [104], on the other hand, was formulated specifically with the speech enhancement application in mind. While speech enhancement systems are able to reduce the background noise level in a noisy speech signal, they also tend to introduce processing artefacts in the speech signal component and in the background noise, especially as the input SNR decreases. When rating the overall quality of the enhanced signal, different test subjects may put different emphasis on these artefacts. This introduces additional variance in the subjects' ratings leading to reduced reliability of the ratings. The underlying idea of the ITU-T P.835 standard is to reduce this variance by guiding the focus of the subjects on the various processing artefacts. Specifically, the ITU-T P.835 standard asks the subjects to evaluate three aspects of the enhanced signals one at a time, namely, the *speech signal*, the *background noise*, and the *overall effect* of speech and background. The subjects are presented with three copies in succession of a given enhanced (or noisy, unprocessed) speech signal. For the first copy, they are asked to attend only to the speech component and rate it according to the five-category distortion scale shown in Table 8.1(a). For the second copy, they are asked to attend only to the noise background, and rate it according to the five-category intrusiveness scale shown in Table 8.1(b). Finally, the subjects should rate the overall-quality of the third copy according to the five-category quality scale shown in Table 8.1(c), which is the scale used in MOS experiments.

8.1.2 INSTRUMENTAL TEST METHODS

During the typical process of developing a speech enhancement system, formal quality tests such as P.835 cannot be conducted more than once or twice. Therefore, formal tests are supplemented by smaller *informal* listening experiments and instrumental distortion measures. A wide range of such distortion measures exists, see e.g., [101, Chap. 10] and [100] and the references therein. We highlight in this section two of them, namely, the segmental-SNR (SEG-SNR), which is simple to compute and often reported in literature, and the Perceptual Evaluation of Speech Quality (PESQ) measure [105], which shows good correlation with perceived quality.

Table 8.1: P.835 Overall rating scales for rating Table 8.1(a) the speech signal, Table 8.1(b) the background noise, and Table 8.1(c) over-all quality.

Rating	Description
5	Not distorted
4	Slightly distorted
3	Somewhat distorted
2	Fairly distorted
1	Very distorted

(a)

Rating	Description
5	Not noticeable
4	Slightly noticeable
3	Noticeable but not intrusive
2	Somewhat intrusive
1	Very intrusive

(b)

Rating	Description
5	Excellent
4	Good
3	Fair
2	Poor
1	Bad

(c)

The SEG-SNR is defined as

$$\text{SEG-SNR} = \frac{1}{L} \sum_{\ell \in \mathcal{L}} 10 \log_{10} \left(\frac{\|\underline{s}_\ell\|^2}{\|\underline{s}_\ell - \widehat{\underline{s}}_\ell\|^2} \right) \qquad [\text{dB}], \tag{8.1}$$

where $\mathcal{L}$ is the index set of speech-active signal frames, L is the cardinality of $\mathcal{L}$, and $\|\cdot\|$ is the vector 2-norm. Since SEG-SNR compares time domain waveforms, it is easy to create situations where it would fail completely. For example, setting $\widehat{\underline{s}}_\ell = -\underline{s}_\ell, \forall \ell$, the estimated signal would be perceptually indistinguishable from the clean signal, but SEG-SNR would be as low as -6 dB. Although SEG-SNR only correlates fairly well with perceived quality [106], this measure is often reported in literature.

The PESQ measure, which is in fact an ITU standard [105], was originally developed for quality evaluation of speech codec distortions (these distortions are generally much smaller than those encountered in output signals of single channel enhancement systems). Nevertheless, over the last decade, the use of PESQ for evaluating enhanced speech has increased dramatically and is now often reported in literature. For a given (clean, enhanced) speech signal pair, PESQ outputs a single scalar in the range -0.5 to 4.5, which correlates well with perceived quality [106]. Although there exist modified versions of PESQ which may perform slightly better than standard PESQ (see e.g., [106]), our informal listening experiments suggest that standard PESQ is useful as it is robust to changing noise and processing types, particularly at medium to high input SNRs (say, SNR > 0 dB). For lower SNRs, though, PESQ is less reliable.

8.2 EVALUATING INTELLIGIBILITY OF ENHANCED SPEECH

8.2.1 LISTENING TESTS

Generally speaking, speech enhancement systems must produce enhanced signals which are at least as intelligible as the noisy input signal: few, if any, applications exist where a quality improvement achieved at the price of an intelligibility reduction can be tolerated. As most single-channel noise reduction systems tend not to improve the intelligibility, the *de facto* goal of evaluating the intelligibility is to verify that the processing did not decrease it.

As for speech quality, intelligibility is measured via listening tests, and many of the existing tests have their origin in the area of speech coding. Subjects are presented with noisy and/or enhanced speech material and the percentage of phonemes or words they are able to correctly identify is taken as a measure of the intelligibility. Intelligibility tests come in several different variants, which may roughly be characterized by a) whether they are adaptive, e.g., the input SNR is changed as a function of the subjects response, or fixed, i.e., the course of the listening test is independent of the subject responses, b) the type and speech material they use, e.g., nonsense-syllables, words, or entire sentences, c) whether they are open- or closed-set tests, i.e., whether the subject knows the complete set of candidate words in a word test (closed-set), or has no such knowledge (open-set). Defining the intelligibility as the percentage of identified phonemes/words is simple, and listening tests often use this definition when measuring the performance of speech enhancement systems, e.g., [1, 2].

However, using this definition sometimes complicates a reliable measurement. Some of the complications arise because one has to choose, before conducting the test, the SNR of the noisy input signal. Figure 8.1 shows the typical shape of a performance (% correct) vs. SNR curve, also known as a *psychometric* function. From this figure it is clear that if an input SNR is chosen, where performance is close to 100%, then even a large performance difference between two enhancement systems under study may be manifested only in a very small difference in measured performance (this phenomenon is referred to as a *ceiling effect*).

Alternatively, intelligibility can be defined in terms of the *speech reception threshold* (SRT). In a speech enhancement context, this is defined as the input SNR, at which a certain percentage, often 50%, of the phonemes/words in a test sentence was understood. Thus, if a speech enhancement system succeeds in lowering the SRT with respect to the noisy, unprocessed input signals, one may claim that the intelligibility is improved.

Several test methodologies exist, e.g., [107, 108], which are capable of estimating the SRT. The Hearing In Noise Test (HINT) [108] is an SNR adaptive procedure which presents simple sentences to the subjects in an open-set response format. The adaptive procedure lowers the input SNR of successive sentences in pre-defined steps as long as the subject is able to identify the full sentence. When an SNR is reached where the subject is no longer able to repeat all words correctly, the SNR is increased in steps until the subject is again able to identify all words. By repeating this *up-down procedure* and keeping track of the SNRs used, it is possible to derive an accurate estimate of the SNR at which the subject identifies, e.g., 50% of the utterances, see [109] for details. For

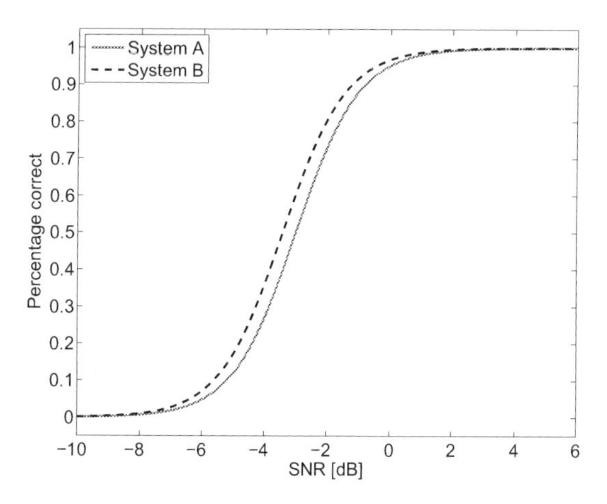

Figure 8.1: General shape of psychometric functions for two enhancement systems, A and B. The general shape is a function of the enhancement systems under study, the speech and noise material used in the test, and the test type.

normal-hearing subjects and speech shaped noise, the SRT resulting from the HINT test is around -2.6 dB for American English [110]. For other adaptive tests, e.g., the semi-closed test proposed by Hagerman, the corresponding SRT is as low as approximately -8 dB [107]; this makes this test less suited for single-channel speech enhancement evaluation, since most enhancement systems tend to have little effect at such a low SNR.

8.2.2 INSTRUMENTAL TEST METHODS

The desire to reduce the number of expensive intelligibility listening tests has spawned interest in developing algorithms for predicting the intelligibility of a given processed speech signal (see e.g., [111] and the references therein). For brevity, we review in the following Taal's Short-Time Objective Intelligibility (STOI) measure [112], as it is simple to implement and use, and shows very high correlation with intelligibility for several different processing conditions.

The STOI measure takes as input a processed noisy speech signal, and the underlying noise-free speech signal. The signals are filtered in one-third octave sub bands to crudely simulate the filtering properties of the cochlea (inner ear), and down sampled. After removing silent regions, which cannot contribute to intelligibility, the temporal envelopes of the sub band signals are normalized and clipped, if necessary. Then, for each time-frequency unit, a short-time sample correlation coefficient is computed based on temporal averages across roughly 300 ms of the temporal envelopes. The final scalar output, $-1 \leq d_{stoi} \leq 1$, of the STOI algorithm is computed as the arithmetic average of the short-time-correlation coefficients across time and sub band indices. Ideally, d_{stoi} is monotonically related to the average intelligibility of the sentence in question, and, indeed, this has been validated for noisy speech enhanced by binary and continuous gain functions, as well as unprocessed noisy speech (see [112]).

CHAPTER 9

Simulation Experiments with Single-Channel Enhancement Systems

In this section we evaluate the algorithms described in the previous sections for enhancement of real speech signals contaminated by noise. Obviously, an exhaustive comparison of all combinations of sub-algorithms of a speech enhancement system is not possible here. For a comparison of sub-algorithms, such as noise PSD estimators and SPP estimators we refer to the previous sections and the references therein. Instead, in this section we demonstrate the general impact of the bigger building blocks in Figure 2.1, (i.e., noise and speech PSD estimation, speech presence estimation, and speech DFT estimation) on speech enhancement performance, as measured by instrumental distortion measures. Specifically, we limit the comparison by considering a base line system and then extend it with algorithms from the different classes. The main goal is to study the relative impact of various building blocks on speech enhancement performance, and not to highlight a particular instance of a building block or algorithm as being better than others.

Table 9.1 gives an overview of the different algorithm combinations that will be included in the comparison. The baseline system consists of the MDFT estimator based on the assumption that speech and noise DFT coefficients have a complex-Gaussian distribution as proposed by Ephraim and Malah in [14]. This estimator can be computed from Eq. (4.5) using the speech magnitude DFT distribution in Eq. (3.2) with $\gamma = 2$ and $\nu = 1$. The noise PSD in this algorithm is estimated using minimum statistics [38] described in Sec. 6.2, while the speech PSD is estimated using the decision-directed approach [14] described in Sec. 7.1. This baseline algorithm does not use a speech presence estimator. Then, in order to demonstrate the influence of a super-Gaussian based estimator, Algorithm 2 makes use of a super-Gaussian based-speech DFT estimator that was proposed in [28]. This estimator can be computed using Eq. (4.5) by setting $\gamma = 1$ and $\nu = 0.6$ in Eq. (3.2). To investigate the impact of low delay noise PSD tracking, Algorithm 3 is based on the SPP-based noise PSD estimation algorithm presented in [42], as discussed in Sec. 6.3. To evaluate the effect of including speech presence probabilities in the speech DFT estimator, Algorithm 4 is equipped with the SPP estimator presented in [88], which is used as indicated in Eq. (5.2). Finally, in order to evaluate the effect of an improved speech PSD estimation on the different performance metrics, Algorithm 5 estimates the speech PSD based on TCS presented in [98] with the bias correction presented in [91], which is discussed in Sec. 7.3. In Algorithm 5, the SPP is turned off to better assess the difference between decision-directed and TCS-based speech PSD estimation.

All algorithms are evaluated using the three instrumental measures discussed in Sec. 8, i.e., PESQ, and segmental SNR as quality measures, and STOI as intelligibility measure. The segmental SNR is evaluated by choosing the index set $\mathcal{L}$ in Eq. (8.1) such that only the time-frames are included where clean speech energy is within the range of 35 dB from the time-frame with maximum clean speech energy. The outcomes d_{stoi}, with $-1 \leq d_{stoi} \leq 1$, of the intelligibility prediction measure STOI are mapped to intelligibility scores using a logistic function, as explained in [112], such that we report the average intelligibility of the sentence in question. The used speech material consists of the complete test-set of the Timit-database [67] with a total duration of 87 minutes, sampled at 16 kHz. The sentences are concatenated per speaker, with an average length of 33 seconds and a minimum and maximum length of approximately 23 and 45 seconds, respectively. The clean speech signals are degraded by modulated speech shaped white noise, where the modulation frequency increases linearly from zero Hz at a rate of 0.5 Hz per 25 seconds. Figure 9.1 shows an example waveform of this noise source. The used (global) input SNRs range from -5 dB to 20 dB in steps of 5 dB. All algorithms make use of 32 ms time-frames taken with 50% overlap with square-root-Hann analysis and synthesis windows. The maximum suppression of all gain functions used as speech DFT estimator is set to -20 dB.

Figure 9.2 shows the performance of the five algorithms in terms of predicted intelligibility improvement measured by STOI, PESQ improvement, and improvement in segmental SNR.

Considering the speech DFT estimator, i.e., Algorithms 1 vs. 2, it is clear that going from a Gaussian-based MDFT estimator to a super-Gaussian-based estimator, both the quality measures as well as the intelligibility measure STOI, show an improvement. Notice that for higher input SNRs, the improvement in terms of STOI, tends to go to zero as for these SNRs all speech material is predicted to be fully intelligible. Further, notice that Algorithm 1 is predicted to lead to a decrease in intelligibility for low input SNRs. The perceptual difference between Algorithms 1 and 2 is mainly perceived as a better noise reduction for the super-Gaussian based estimator, albeit at the cost of somewhat more musical sounding residual noise.

The influence of a low delay noise PSD estimator becomes clear when comparing Algorithms 2 and 3 when the noise source becomes non-stationary. The MS approach in Algorithm 2 is not able to follow the quickly increasing noise levels in the noise source, leading to a lot of residual noise compared to the faster SPP-based noise tracker in Algorithm 3. As this noise PSD tracker yields an improved estimate of the increasing noise level, a better noise reduction is obtained. At the same time, this means that more suppression is applied, which automatically results in somewhat more suppression of the speech signal when the noise level increases to higher levels. This is reflected by the objective performance measures in Figure 9.2 by a much larger PESQ and segmental SNR improvement and a smaller improvement in terms of intelligibility for Algorithm 3 compared to Algorithm 2. The slightly lower predicted intelligibility improvement might be caused by the fact that due to the better noise PSD estimate of increasing noise levels, automatically more speech suppression

Table 9.1: Algorithm specifications used in evaluation.

Alg. number	speech DFT (Sec. 4)	noise PSD (Sec. 6)	speech PSD (Sec. 7)	SPP (Sec. 5)
1	MDFT - Gaussian [14]	MS-based [38]	decision-directed [14]	No
2	MDFT - super-Gaussian [28]	MS-based [38]	decision-directed [14]	No
3	MDFT - super-Gaussian [28]	SPP-based [42]	decision-directed [14]	No
4	MDFT - super-Gaussian [28]	SPP-based [42]	decision-directed [14]	yes - TCS [88]
5	MDFT - super-Gaussian [28]	SPP-based [42]	TCS [98] with [91]	No

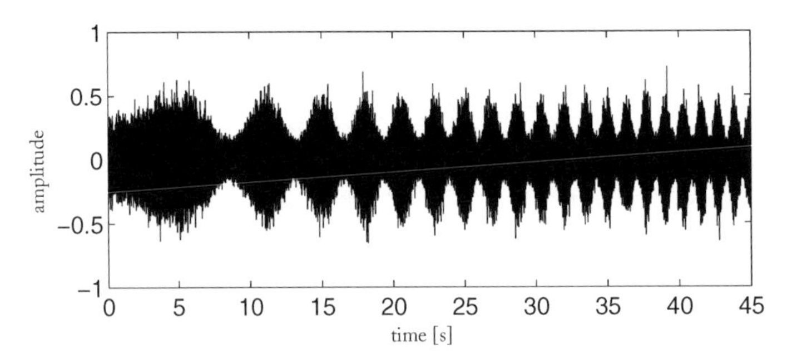

Figure 9.1: Waveform of the noise source used in the experiments.

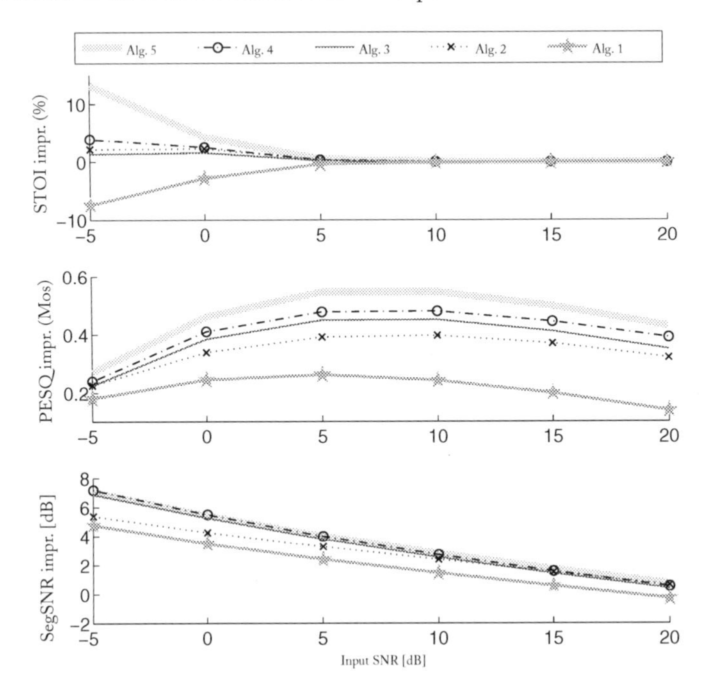

Figure 9.2: Performance evaluation in terms of improvement over the noisy speech of STOI, PESQ, and segmental SNR.

will be applied. This could be overcome by developing speech DFT estimators under distortion measures that better reflect speech intelligibility and take this into account. During intervals where the noise source remains stationary, both Algorithms 2 and 3 give a similar perceptual impression.

The SPP incorporated in Algorithm 4 leads to somewhat lower residual noise than when no SPP is used, visible in terms of slightly better PESQ and STOI improvement. Furthermore, the SPP estimator results in an effective reduction of musical noise. However, the amount of musical noise is reflected only very indirectly by the instrumental measures. Notice that when using a SPP estimator, special care needs to be taken to overcome an over-suppression of speech due to the multiplicative use of the SPP.

Finally, the improved speech PSD estimator based on TCS in Algorithm 5 leads to smoother estimates of the speech PSD resulting in a much lower amount of musical noise. At the same time, TCS results in a better preservation of the speech spectral structure as compared to the commonly used decision-directed approach. This preservation of the speech spectral structure results in less speech distortions in a speech enhancement task. That the benefits of TCS are perceptually relevant is best reflected by the PESQ and STOI measures, which predict an improved quality and intelligibility if TCS is used for speech PSD estimation.

CHAPTER 10

Future Directions

Single channel noise reduction systems for speech enhancement are ubiquitous, either as stand-alone algorithms or as parts of multi-microphone systems, namely, as post-processing steps for beamforming algorithms. The range of applications where multi-microphone systems may be successfully used can be expected to increase in the future, as hardware costs, e.g., for microphones and signal processing computational power decreases, and battery technology improves. In this way, applications where stand-alone single-microphone speech enhancement systems dominate today may be replaced by multi-microphone systems, and we therefore expect increased focus on single-channel systems used as post-processing steps. However, many applications rule out multi-microphone solutions, primarily due to limitations in possible locations of the microphones, e.g., hearing aid applications, certain forensic applications, etc. So, stand-alone single-microphone enhancement systems will remain in focus as well.

Over the last few years, we have witnessed a broadening of the range of applications where single-channel speech enhancement algorithms can be successfully used. This is perhaps most notably seen with the recent development of fast noise power spectral density estimators and speech power spectral density estimators, which have improved the robustness of single-channel noise reduction systems to changing acoustical environments. Furthermore, recently single-channel noise reduction systems have been applied in domains which originally were considered outside the scope of the traditional additive-and-independent noise model, e.g., reverberance reduction [6]. We expect this trend to continue. Specifically, with the general expectation that noise reduction systems should work anywhere and at any time without user intervention, we expect that increased focus is directed toward single-channel enhancement systems as parts of larger noise reduction systems which are robust to changing acoustical situations.

As we have outlined in the previous sections, there exist a number of universal tasks that any single-channel noise reduction system faces, essentially independently of the application at hand. These are: target signal PSD estimation, noise PSD estimation, target signal estimation, and, potentially, target presence probability estimation. Research has been active in these areas for several decades, and as demonstrated in the simulation study in Sec. 9, this effort has indeed led to quite significant performance improvements. Specifically, recent progress has been eased by researchers publishing their algorithm implementations online, allowing easy and error-free comparison to state-of-the-art. It is our clear impression that improvements in these areas can still be found, but as these core areas are indeed well-researched, we expect the development to be characterized by small incremental improvements, which, across time, may lead to significantly better systems than today.

An example of an emerging field that achieved less attention in the past is the estimation of the clean speech spectral phase [61, 62]. Recent results show that an estimate of the clean speech phase can also be employed to make spectral amplitude estimation more robust [64], thus pushing the limits of single channel speech enhancement algorithms further.

A new and important challenge in the coming decade may be found in the development of speech enhancement systems which produce enhanced signals that are optimal as measured by a model of the human auditory system/instrumental measure rather than optimal in some MMSE sense. This idea is by no means new in the single channel speech enhancement community, but the auditory models/instrumental measures used so far have proven either mathematically intractable and/or incapable of predicting the impact on quality or intelligibility of a given processing strategy. However, advances have been made in our understanding of the human auditory system, both in terms of the auditory periphery, but also in terms of higher stages of the auditory system, and recently, simple computational models or instrumental measures have emerged, which are both relatively mathematically tractable and yet successful in predicting quality or intelligibility aspects of a given processed signal [112]. We believe that letting such models of auditory perception aspects guide the processing applied to the noisy signal may lead to new and improved noise reduction systems in the decade to come.

Bibliography

[1] J. Jensen and R. C. Hendriks, "Spectral magnitude minimum mean-square error estimation using binary and continuous gain functions," *IEEE Trans. Audio, Speech, Language Process.*, vol. 20, no. 1, pp. 92–102, Jan. 2012. DOI: 10.1109/TASL.2011.2157685 1, 46

[2] Y. Hu and P. C. Loizou, "A comparative intelligibility study of single-microphone noise reduction algorithms," *J. Acoust. Soc. Amer.*, vol. 122, no. 3, pp. 1777–1786, Sep. 2007. DOI: 10.1121/1.2766778 1, 46

[3] Y. Hu and P. C. Loizou, "Evaluation of objective quality measures for speech enhancement," *IEEE Trans. Audio, Speech, Language Process.*, vol. 16, no. 1, pp. 229–238, Jan. 2008. DOI: 10.1109/TASL.2007.911054 1

[4] K. U. Simmer, J. Bitzer, and C. Marro, "Post-filtering techniques," in *Microphone Arrays: Signal Processing Techniques and Applications*, M. S. Brandstein and C. Ward, Eds. Springer, Berlin, 2001, pp. 39–60. 2

[5] R. C. Hendriks, R. Heusdens, U. Kjems, and J. Jensen, "On optimal multi-channel mean-squared error estimators for speech enhancement," *IEEE Signal Process. Lett.*, vol. 16, no. 10, pp. 885–888, Oct. 2009. DOI: 10.1109/LSP.2009.2026205 2

[6] H. W. Löllmann and P. Vary, "A blind speech enhancement algorithm for the suppression of late reverberation and noise," in *IEEE Int. Conf. Acoust., Speech, Signal Process. (ICASSP)*, Taipei, Taiwan, Apr. 2009, pp. 3989–3992. DOI: 10.1109/ICASSP.2009.4960502 5, 55

[7] M. Dendrinos, S. Bakamides, and G. Carayannis, "Speech enhancement from noise: A regenerative approach," *ELSEVIER Speech Commun.*, vol. 10, no. 2, pp. 45–57, Feb. 1991. DOI: 10.1016/0167-6393(91)90027-Q 6

[8] S. H. Jensen, P. C. Hansen, S. D. Hansen, and J. A. Sørensen, "Reduction of Broad-Band Noise in Speech by Truncated QSVD," *IEEE Trans. Speech Audio Process.*, vol. 3, no. 6, pp. 439–448, Nov. 1995. DOI: 10.1109/89.482211 6

[9] Y. Ephraim and H. L. V. Trees, "A signal subspace approach for speech enhancement," *IEEE Trans. Speech Audio Process.*, vol. 3, no. 4, pp. 251–266, Jul. 1995. DOI: 10.1109/89.397090 6, 11

[10] F. Jabloun and B. Champagne, "Incorporating the human hearing properties in the signal subspace approach for speech enhancement," *IEEE Trans. Speech Audio Process.*, vol. 11, no. 6, pp. 700–708, Nov. 2003. DOI: 10.1109/TSA.2003.818031 6, 11

[11] J. Jensen and R. Heusdens, "Improved subspace-based single-channel speech enhancement using super-Gaussian priors," *IEEE Trans. Audio, Speech, Language Process.*, vol. 15, no. 3, pp. 862–872, Mar. 2007. DOI: 10.1109/TASL.2005.885939 6, 8

[12] S. F. Boll, "Suppression of acoustic noise in speech using spectral subtraction," *IEEE Trans. Acoust., Speech, Signal Process.*, vol. ASSP-27, no. 2, pp. 113–120, Apr. 1979. DOI: 10.1109/TASSP.1979.1163209 6, 8, 16

[13] M. Berouti, R. Schwartz, and J. Makhoul, "Enhancement of speech corrupted by acoustic noise," in *IEEE Int. Conf. Acoust., Speech, Signal Process. (ICASSP)*, Washington, DC, USA, Apr. 1979, pp. 208–211. DOI: 10.1109/ICASSP.1979.1170788 6, 8, 16

[14] Y. Ephraim and D. Malah, "Speech enhancement using a minimum mean-square error short-time spectral amplitude estimator," *IEEE Trans. Acoust., Speech, Signal Process.*, vol. 32, no. 6, pp. 1109–1121, Dec. 1984. DOI: 10.1109/TASSP.1984.1164453 6, 9, 10, 15, 19, 21, 23, 24, 25, 28, 37, 38, 49, 51

[15] R. Martin, "Speech enhancement based on minimum mean-square error estimation and supergaussian priors," *IEEE Trans. Speech Audio Process.*, vol. 13, no. 5, pp. 845–856, Sep. 2005. DOI: 10.1109/TSA.2005.851927 6, 8, 10, 15, 19, 20, 22, 25

[16] I. Cohen, "Relaxed statistical model for speech enhancement and a priori SNR estimation," *IEEE Trans. Speech Audio Process.*, vol. 13, no. 5, pp. 870–881, Sep. 2005. DOI: 10.1109/TSA.2005.851940 6, 9, 38, 39

[17] J. S. Lim and A. V. Oppenheim, "Enhancement and bandwidth compression of noisy speech," *Proc. of the IEEE*, vol. 67, no. 12, pp. 1586–1604, Dec. 1979. DOI: 10.1109/PROC.1979.11540 6, 9, 15, 16, 18

[18] N. Virag, "Single channel speech enhancement based on masking properties of the human auditory system," *IEEE Trans. Speech Audio Process.*, vol. 7, no. 2, pp. 126–137, Mar. 1999. DOI: 10.1109/89.748118 6, 11

[19] S. Gazor and W. Zhang, "Speech enhancement employing Laplacian-Gaussian mixture," *IEEE Trans. Speech Audio Process.*, vol. 13, no. 5, pp. 896–904, Sep. 2005. DOI: 10.1109/TSA.2005.851943 6, 8

[20] T. Gülzow, A. Engelsberg, and U. Heute, "Comparison of a discrete wavelet transformation and a nonuniform polyphase filterbank applied to spectral-subtraction speech enhancement," *ELSEVIER Signal Process.*, vol. 64, no. 1, pp. 5–19, Jan. 1998. DOI: 10.1016/S0165-1684(97)00172-2 6

[21] Y. Ephraim, "Statistical-model-based speech enhancement systems," *Proc. IEEE*, vol. 80, no. 10, pp. 1526–1555, Oct. 1992. DOI: 10.1109/5.168664 6, 9

[22] R. J. McAulay and M. L. Malpass, "Speech enhancement using a soft-decision noise suppression filter," *IEEE Trans. Acoust., Speech, Signal Process.*, vol. 28, no. 2, pp. 137–145, Apr. 1980. DOI: 10.1109/TASSP.1980.1163394 8, 13, 15, 16, 25, 28

[23] J. Jensen, I. Batina, R. C. Hendriks, and R. Heusdens, "A study of the distribution of time-domain speech samples and discrete Fourier coefficients," in *Proc. IEEE First BENELUX/DSP Valley Signal Processing Symposium*, Antwerp, Belgium, Apr. 2005, pp. 155–158. 8

[24] T. Lotter and P. Vary, "Speech enhancement by MAP spectral amplitude estimation using a super-Gaussian speech model," *EURASIP J. Applied Signal Process.*, vol. 2005, no. 7, pp. 1110–1126, Jan. 2005. DOI: 10.1155/ASP.2005.1110 8, 13, 15, 19

[25] T. Lotter and P. Vary, "Noise reduction by maximum a posteriori spectral amplitude estimation with supergaussian speech modeling," in *Int. Workshop Acoustic Echo, Noise Control (IWAENC)*, Kyoto, Japan, Sep. 2003, pp. 83–86. 8

[26] I. Cohen, "Supergaussian garch models for speech signals," in *ISCA Interspeech*, Lisbon, Portugal, Sep. 2005, pp. 2053–2056. 8

[27] I. Andrianakis and P. R. White, "MMSE speech spectral amplitude estimators with Chi and Gamma speech priors," in *IEEE Int. Conf. Acoust., Speech, Signal Process. (ICASSP)*, Toulouse, France, May 2006, pp. 1068–1071. DOI: 10.1109/ICASSP.2006.1660842 8, 14, 19, 21

[28] J. S. Erkelens, R. C. Hendriks, R. Heusdens, and J. Jensen, "Minimum mean-square error estimation of discrete Fourier coefficients with generalized Gamma priors," *IEEE Trans. Audio, Speech, Language Process.*, vol. 15, no. 6, pp. 1741–1752, Aug. 2007. DOI: 10.1109/TASL.2007.899233 8, 14, 19, 20, 21, 22, 49, 51

[29] J. H. L. Hansen and M. A. Clements, "Constrained iterative speech enhancement with application to speech recognition," *IEEE Trans. Signal Process.*, vol. 39, no. 4, pp. 795–805, Apr. 1991. DOI: 10.1109/78.80901 9

[30] T. V. Sreenivas and P. Kirnapure, "Codebook constrained Wiener filtering for speech enhancement," *IEEE Trans. Speech Audio Process.*, vol. 4, no. 5, pp. 383–389, Sep. 1996. DOI: 10.1109/89.536932 9

[31] C. Breithaupt, T. Gerkmann, and R. Martin, "Cepstral smoothing of spectral filter gains for speech enhancement without musical noise," *IEEE Signal Process. Lett.*, vol. 14, no. 12, pp. 1036–1039, Dec. 2007. DOI: 10.1109/LSP.2007.906208 9

[32] S. Srinivasan, J. Samuelsson, and W. B. Kleijn, "Codebook-based Bayesian speech enhancement for nonstationary environments," *IEEE Trans. Audio, Speech, Language Process.*, vol. 15, pp. 441–452, Feb. 2007. DOI: 10.1109/TASL.2006.881696 9

[33] I. Cohen, "Modeling speech signals in the time-frequency domain using GARCH," *ELSEVIER Signal Process.*, vol. 84, no. 12, pp. 2453–2459, Dec. 2004. DOI: 10.1016/j.sigpro.2004.09.001 9

[34] I. Cohen, "Speech spectral modeling and enhancement based on autoregressive conditional heteroscedasticity models," *ELSEVIER Signal Process.*, vol. 86, no. 4, pp. 698–709, Apr. 2006. DOI: 10.1016/j.sigpro.2005.06.005 9, 38, 39

[35] X. Xiao and R. M. Nickel, "Speech enhancement with inventory style speech resynthesis," *IEEE Trans. Audio, Speech, Language Process.*, vol. 18, no. 6, pp. 1243–1257, Aug. 2010. DOI: 10.1109/TASL.2009.2031793 9

[36] J. Ming, R. Srinivasan, and D. Crookes, "A corpus-based approach to speech enhancement from nonstationary noise," *IEEE Trans. Audio, Speech, Language Process.*, vol. 19, no. 4, pp. 822–836, May 2011. DOI: 10.1109/TASL.2010.2064312 9

[37] Y. Ephraim and D. Malah, "Speech enhancement using a minimum mean-square error log-spectral amplitude estimator," *IEEE Trans. Acoust., Speech, Signal Process.*, vol. 33, no. 2, pp. 443–445, Apr. 1985. DOI: 10.1109/TASSP.1985.1164550 10, 21, 39

[38] R. Martin, "Noise power spectral density estimation based on optimal smoothing and minimum statistics," *IEEE Trans. Speech Audio Process.*, vol. 9, no. 5, pp. 504–512, Jul. 2001. DOI: 10.1109/89.928915 10, 30, 31, 33, 34, 35, 49, 51

[39] I. Cohen, "Noise spectrum estimation in adverse environments: Improved minima controlled recursive averaging," *IEEE Trans. Speech Audio Process.*, vol. 11, no. 5, pp. 466–475, Sep. 2003. DOI: 10.1109/TSA.2003.811544 10, 31, 33, 34

[40] S. Rangachari and P. C. Loizou, "A noise-estimation algorithm for highly non-stationary environments," *ELSEVIER Speech Commun.*, vol. 48, no. 2, pp. 220–231, Feb. 2006. DOI: 10.1016/j.specom.2005.08.005 10, 33

[41] R. C. Hendriks, R. Heusdens, and J. Jensen, "MMSE based noise PSD tracking with low complexity," in *IEEE Int. Conf. Acoust., Speech, Signal Process. (ICASSP)*, Dallas, TX, USA, Mar. 2010, pp. 4266–4269. DOI: 10.1109/ICASSP.2010.5495680 10, 31, 35

[42] T. Gerkmann and R. C. Hendriks, "Unbiased MMSE-based noise power estimation with low complexity and low tracking delay," *IEEE Trans. Audio, Speech, Language Process.*, vol. 20, no. 4, pp. 1383–1393, May 2012. DOI: 10.1109/TASL.2011.2180896 10, 31, 33, 34, 35, 49, 51

[43] R. C. Hendriks, J. Jensen, and R. Heusdens, "Noise tracking using DFT domain subspace decompositions," *IEEE Trans. Audio, Speech, Language Process.*, vol. 16, no. 3, pp. 541–553, Mar. 2008. DOI: 10.1109/TASL.2007.914977 10, 31, 35

[44] D. Ealey, H. Kelleher, and D. Pearce, "Harmonic tunnelling: tracking non-stationary noises during speech," in *ISCA Eurospeech*, Budapest, Hungary, Sep. 1999, pp. 437–440. 10

[45] P. C. Loizou, "Speech enhancement based on perceptually motivated Bayesian estimators of the magnitude spectrum," *IEEE Trans. Speech Audio Process.*, vol. 13, no. 5, pp. 857–869, Sep. 2005. DOI: 10.1109/TSA.2005.851929 10, 19, 21

[46] C. H. You, S. N. Koh, and S. Rahardja, "β-order MMSE spectral amplitude estimation for speech enhancement," *IEEE Trans. Speech Audio Process.*, vol. 13, no. 4, pp. 475–486, Jul. 2005. DOI: 10.1109/TSA.2005.848883 10, 19, 21

[47] C. Breithaupt, M. Krawczyk, and R. Martin, "Parameterized MMSE spectral magnitude estimation for the enhancement of noisy speech," in *IEEE Int. Conf. Acoust., Speech, Signal Process. (ICASSP)*, Las Vegas, NV, USA, Apr. 2008, pp. 4037–4040. DOI: 10.1109/ICASSP.2008.4518540 10, 14, 19, 21

[48] C. Breithaupt and R. Martin, "Analysis of the decision-directed SNR estimator for speech enhancement with respect to low-SNR and transient conditions," *IEEE Trans. Audio, Speech, Language Process.*, vol. 19, no. 2, pp. 277–289, Feb. 2011. DOI: 10.1109/TASL.2010.2047681 10, 14, 19, 21, 25, 38

[49] ISO/MPEG Committee, *Coding of moving pictures and associated audio for digital storage media at up to about 1.5Mbit/s — part 3: Audio*, ISO/IEC 11172-3, 1993. 11

[50] D. E. Tsoukalas, J. N. Mourjopoulos, and G. Kokkinakis, "Speech enhancement based on audible noise suppression," *IEEE Trans. Speech Audio Process.*, vol. 5, no. 6, pp. 497–514, Nov. 1997. DOI: 10.1109/89.641296 11

[51] Y. Hu and P. C. Loizou, "A perceptually motivated approach for speech enhancement," *IEEE Trans. Speech Audio Process.*, vol. 11, no. 5, pp. 457–465, Sep. 2003. DOI: 10.1109/TSA.2003.815936 11

[52] S. Gustafsson, R. Martin, P. Jax, and P. Vary, "A psychoacoustic approach to combined acoustic echo cancellation and noise reduction," *IEEE Trans. Speech Audio Process.*, vol. 10, no. 5, pp. 245–256, Jul. 2002. DOI: 10.1109/TSA.2002.800553 11

[53] S. Jo and C. D. Yoo, "Psychoacoustically constrained and distortion minimized speech enhancement," *IEEE Trans. Audio, Speech, Language Process.*, vol. 18, no. 8, pp. 2099–2110, Nov. 2010. DOI: 10.1109/TASL.2010.2041119 11

[54] K. Paliwal, K. Wojcicki, and B. Schwerin, "Single-channel speech enhancement using spectral subtraction in the short-time modulation domain," *ELSEVIER Speech Commun.*, vol. 52, pp. 450–475, May 2010. DOI: 10.1016/j.specom.2010.02.004 11

[55] J. Schnupp, I. Nelken, and A. King, *Auditory Neuroscience: making sense of sound.* Massachusetts Institute of Technology, 2011. 11

[56] B. Kollmeier and R. Koch, "Speech enhancement based on physiological and psychoacoustical models of modulation perception and binaural interaction." *J. Acoust. Soc. Amer.*, vol. 95, no. 3, pp. 1593–602, Mar. 1994. DOI: 10.1121/1.408546 11

[57] R. Drullman, J. Festen, and R. Plomp, "Effect of reducing slow temporal modulations on speech reception." *J. Acoust. Soc. Amer.*, vol. 95, no. 5, pp. 2670–2680, May 1994. DOI: 10.1121/1.409836 11

[58] R. Drullman, J. Festen, and R. Plomp, "Effect of temporal envelope smearing on speech reception." *J. Acoust. Soc. Amer.*, vol. 95, no. 2, pp. 1053–1064, Feb. 1994. DOI: 10.1121/1.408467 11

[59] D. R. Brillinger, *Time Series: Data Analysis and Theory.* San Francisco, CA, USA: Holden-Day, 1981. 13

[60] J. S. Erkelens, R. C. Hendriks, and R. Heusdens, "On the estimation of complex speech DFT coefficients without assuming independent real and imaginary parts," *IEEE Signal Process. Lett.*, vol. 15, pp. 213–216, 2008. DOI: 10.1109/LSP.2007.911730 15, 20, 22

[61] K. Paliwal, K. Wójcicki, and B. Shannon, "The importance of phase in speech enhancement," *ELSEVIER Speech Commun.*, vol. 53, no. 4, pp. 465–494, Apr. 2011. DOI: 10.1016/j.specom.2010.12.003 15, 56

[62] T. Gerkmann, M. Krawczyk, and R. Rehr, "Phase estimation in speech enhancement — unimportant, important, or impossible?" in *27th Convention of Electrical and Electronics Engineers in Israel*, Eilat, Israel, Nov. 2012. DOI: 10.1109/EEEI.2012.6376931 15, 56

[63] M. Krawczyk and T. Gerkmann, "STFT phase improvement for single channel speech enhancement," in *Int. Workshop Acoustic Echo, Noise Control (IWAENC)*, Aachen, Germany, Sep. 2012. 15

[64] T. Gerkmann and M. Krawczyk, "MMSE-optimal spectral amplitude estimation given the STFT-phase," *IEEE Signal Process. Lett.*, vol. 20, no. 2, Feb. 2013. DOI: 10.1109/LSP.2012.2233470 15, 56

[65] I. Tashev and A. Acero, "Statistical modeling of the speech signal," in *Int. Workshop Acoustic Echo, Noise Control (IWAENC)*, Tel Aviv, Israel, Aug. 2010. 15

[66] T. Gerkmann and R. Martin, "Empirical distributions of DFT-domain speech coefficients based on estimated speech variances," in *Int. Workshop Acoustic Echo, Noise Control (IWAENC)*, Tel Aviv, Israel, Aug. 2010. 15

[67] J. S. Garofolo, "DARPA TIMIT acoustic-phonetic speech database," *National Institute of Standards and Technology (NIST)*, 1988. 15, 29, 50

[68] N. Wiener, *Extrapolation, Interpolation and Smoothing of Stationary Time Series: With Engineering Applications*, principles of electrical engineering series ed. MIT Press, 1949. 18, 22

[69] R. C. Hendriks and R. Heusdens, "On linear versus non-linear magnitude-DFT estimators and the influence of super-Gaussian speech priors," in *IEEE Int. Conf. Acoust., Speech, Signal Process. (ICASSP)*, Dallas, TX, USA, Mar. 2010, pp. 4750 –4753. DOI: 10.1109/ICASSP.2010.5495172 18

[70] D. H. Brandwood, "A complex gradient operator and its application in adaptive array theory," *IEE Proc.*, vol. 130, pts. F and H, no. 1, pp. 11 –16, Feb. 1983. DOI: 10.1049/ip-f-1:19830003 18

[71] C. W. Therrien, *Discrete Random Signals and Statistical Signal Processing*. Englewood Cliffs, NJ: Prentice-Hall, 1992. 18, 19

[72] P. J. Wolfe and S. J. Godsill, "Simple alternatives to the Ephraim and Malah suppression rule for speech enhancement," in *IEEE Workshop Stat. Signal Process.*, Singapore, Aug. 2001, pp. 496–499. DOI: 10.1109/SSP.2001.955331 19, 21, 39

[73] R. Martin, "Speech enhancement using MMSE short time spectral estimation with Gamma distributed speech priors," in *IEEE Int. Conf. Acoust., Speech, Signal Process. (ICASSP)*, Orlando, FL, USA, May 2002, pp. 253–256. DOI: 10.1109/ICASSP.2002.5743702 19, 22

[74] R. Martin and C. Breithaupt, "Speech enhancement in the DFT domain using Laplacian speech priors," in *Int. Workshop Acoustic Echo, Noise Control (IWAENC)*, Kyoto, Japan, Sep. 2003, pp. 87–90. 19, 22

[75] I. Andrianakis and P. R. White, "Speech spectral amplitude estimators using optimally shaped Gamma and Chi priors," *ELSEVIER Speech Commun.*, vol. 51, no. 1, pp. 1–14, Jan. 2009. DOI: 10.1016/j.specom.2008.05.018 19

[76] P. Vary and R. Martin, *Digital Speech Transmission: Enhancement, Coding And Error Concealment*. Chichester, West Sussex, UK: John Wiley & Sons, 2006. DOI: 10.1002/0470031743 19, 24

[77] R. C. Hendriks, J. S. Erkelens, and R. Heusdens, "Comparison of complex-DFT estimators with and without the independence assumption of real and imaginary parts," in *IEEE Int. Conf. Acoust., Speech, Signal Process. (ICASSP)*, Las Vegas, NV, USA, Apr. 2008, pp. 4033–4036. DOI: 10.1109/ICASSP.2008.4518539 20, 22

[78] R. C. Hendriks, J. S. Erkelens, J. Jensen, and R. Heusdens, "Minimum mean-square error amplitude estimators for speech enhancement under the generalized Gamma distribution," in *Int. Workshop Acoustic Echo, Noise Control (IWAENC)*, Paris, France, Sep. 2006. 21

[79] J. S. Erkelens, J. Jensen, and R. Heusdens, "Improved speech spectral variance estimation under the generalized Gamma distribution," in *IEEE BENELUX/DSP Valley Signal Process. Symp.*, Mar. 2007, pp. 43–46. 21

[80] R. C. Hendriks, R. Heusdens, and J. Jensen, "Log-spectral magnitude MMSE estimators under super-Gaussian densities," in *ISCA Interspeech*, Brighton, United Kingdom, Sep. 2009, pp. 1319–1322. 21

[81] E. Plourde and B. Champagne, "Perceptually based speech enhancement using the weighted β-SA estimator," in *IEEE Int. Conf. Acoust., Speech, Signal Process. (ICASSP)*, Las Vegas, NV, USA, Apr. 2008, pp. 4193 –4196. 21

[82] R. Martin, "Statistical methods for the enhancement of noisy speech," in *Int. Workshop Acoustic Echo, Noise Control (IWAENC)*, kyoto, Japan, Sep. 2003, pp. 1–6. DOI: 10.1007/3-540-27489-8_3 22

[83] J. Jensen, R. C. Hendriks, J. S. Erkelens, and R. Heusdens, "MMSE estimation of complex-valued discrete Fourier coefficients with generalized Gamma priors," in *ISCA Interspeech*, Pittsburgh, PA, USA, Sep. 2006, pp. 257–260. 22

[84] I. Cohen and B. Berdugo, "Speech enhancement for non-stationary noise environments," *ELSEVIER Signal Process.*, vol. 81, no. 11, pp. 2403–2418, Nov. 2001. DOI: 10.1016/S0165-1684(01)00128-1 23, 24, 25, 27, 28

[85] D. Malah, R. Cox, and A. Accardi, "Tracking speech-presence uncertainty to improve speech enhancement in non-stationary noise environments," in *IEEE Int. Conf. Acoust., Speech, Signal Process. (ICASSP)*, Phoenix, AZ, USA, Mar. 1999, pp. 789–792. DOI: 10.1109/ICASSP.1999.759789 23, 24, 25, 26, 27, 28

[86] T. Gerkmann, C. Breithaupt, and R. Martin, "Improved a posteriori speech presence probability estimation based on a likelihood ratio with fixed priors," *IEEE Trans. Audio, Speech, Language Process.*, vol. 16, no. 5, pp. 910–919, Jul. 2008. DOI: 10.1109/TASL.2008.921764 24, 25, 26, 28

[87] B. Fodor and T. Fingscheidt, "MMSE speech enhancement under speech presence uncertainty assuming (generalized) Gamma speech priors throughout," in *IEEE Int. Conf. Acoust., Speech, Signal Process. (ICASSP)*, Kyoto, Japan, Mar. 2012, pp. 4033–4036. DOI: 10.1109/ICASSP.2012.6288803 25

[88] T. Gerkmann, M. Krawczyk, and R. Martin, "Speech presence probability estimation based on temporal cepstrum smoothing," in *IEEE Int. Conf. Acoust., Speech, Signal Process. (ICASSP)*, Dallas, TX, USA, Mar. 2010, pp. 4254–4257. DOI: 10.1109/ICASSP.2010.5495677 25, 27, 28, 49, 51

[89] J. Sohn and W. Sung, "A voice activity detector employing soft decision based noise spectrum adaptation," in *IEEE Int. Conf. Acoust., Speech, Signal Process. (ICASSP)*, vol. 1, Seattle, WA, USA, May 1998, pp. 365–368. DOI: 10.1109/ICASSP.1998.674443 25, 33, 34

[90] N. Kim and J. Chang, "Spectral enhancement based on global soft decision," *IEEE Signal Process. Lett.*, vol. 7, no. 5, pp. 108–110, May 2000. DOI: 10.1109/97.841154 28

[91] T. Gerkmann and R. Martin, "On the statistics of spectral amplitudes after variance reduction by temporal cepstrum smoothing and cepstral nulling," *IEEE Trans. Signal Process.*, vol. 57, no. 11, pp. 4165–4174, Nov. 2009. DOI: 10.1109/TSP.2009.2025795 28, 40, 50, 51

[92] J. van Gerven and F. Xie, "A comparative study of speech detection methods," in *European Conference on Speech Communication and Technology*, Sep. 1997, pp. 1095–1098. 29, 30, 31

[93] R. Martin, "Spectral subtraction based on minimum statistics," in *EURASIP Europ. Signal Process. Conf. (EUSIPCO)*, Edinburgh, Scotland, Sep. 1994, pp. 1182–1185. 30

[94] R. Yu, "A low-complexity noise estimation algorithm based on smoothing of noise power estimation and estimation bias correction," in *IEEE Int. Conf. Acoust., Speech, Signal Process. (ICASSP)*, Taipei, Taiwan, Apr. 2009, pp. 4421–4424. DOI: 10.1109/ICASSP.2009.4960610 35

[95] R. C. Hendriks, J. Jensen, and R. Heusdens, "DFT domain subspace based noise tracking for speech enhancement," in *ISCA Interspeech*, Antwerp, Belgium, Aug. 2007, pp. 830–833. 35

[96] J. Taghia, J. Taghia, N. Mohammadiha, J. Sang, V. Bouse, and R. Martin, "An evaluation of noise power spectral density estimation algorithms in adverse acoustic environments," in *IEEE Int. Conf. Acoust., Speech, Signal Process. (ICASSP)*, Dallas, TX, USA, May 2011. DOI: 10.1109/ICASSP.2011.5947389 35

[97] O. Cappé, "Elimination of the musical noise phenomenon with the Ephraim and Malah noise suppressor," *IEEE Trans. Speech Audio Process.*, vol. 2, no. 2, pp. 345–349, Apr. 1994. DOI: 10.1109/89.279283 38

[98] C. Breithaupt, T. Gerkmann, and R. Martin, "A novel a priori SNR estimation approach based on selective cepstro-temporal smoothing," in *IEEE Int. Conf. Acoust., Speech, Signal Process. (ICASSP)*, Las Vegas, NV, USA, Apr. 2008, pp. 4897–4900.
DOI: 10.1109/ICASSP.2008.4518755 39, 40, 49, 51

[99] S. R. Quackenbush, T. P. Barnwell, III, and M. A. Clements, *Objective Measures of Speech Quality*. Englewood Cliffs, NJ, USA: Prentice Hall, 1988. 43

[100] J. R. Deller, J. H. L. Hansen, and J. G. Proakis, *Discrete-Time Processing of Speech Signals*. New York, NY, USA: IEEE Press, 1993. 43, 44

[101] P. C. Loizou, *Speech Enhancement - Theory and Practice*. Boca Raton, FL, USA: CRC Press, Taylor & Francis Group, 2007. 43, 44

[102] International Telecommunication Union — Radio Communication Sector, *Recommendation BS.562-3, Subjective assessment of sound quality*, 1990. 44

[103] W. D. Voiers, "Diagnostic acceptability measure for speech communication systems," in *IEEE Int. Conf. Acoust., Speech, Signal Process. (ICASSP)*, Hartford, CT, USA, May 1977, pp. 204–207. DOI: 10.1109/ICASSP.1977.1170198 44

[104] International Telecommunication Union — Telecommunication Standardization Sector of ITU, *ITU-T Recommendation P.835. Subjective test methodology for evaluating speech communication systems that include noise suppression algorithm*, 2003. 44

[105] ITU-T, "Perceptual evaluation of speech quality (PESQ)," *ITU-T Recommendation P.862*, 2001. 44, 45

[106] P. C. Loizou, "Speech quality assessment," in *Multimedia Analysis, Processing and Communications*, Lin *et al.*, Eds. Springer Verlag, 2011, vol. 346, pp. 623–654. 45

[107] B. Hagerman, "Sentences for testing speech intelligibility in noise," *Scand. Audiol.*, vol. 11, no. 2, pp. 79–87, 1982. DOI: 10.3109/01050398209076203 46, 47

[108] M. Nilsson, S. Soli, and J. Sullivan, "Development of hearing in noise test for the measurement of speech reception thresholds in quiet and in noise," *J. Acoust. Soc. Amer.*, vol. 95, no. 2, pp. 1085–1099, Feb. 1994. DOI: 10.1121/1.408469 46

[109] H. Levitt, "Transformed up-down methods in psychoacoustics," *J. Acoust. Soc. Amer.*, vol. 49, no. 2 (Part 2), pp. 467–477, Feb. 1971. DOI: 10.1121/1.1912375 46

[110] S. D. Soli and L. L. N. Wong, "Assessment of speech intelligibility in noise with the Hearing in Noise Test," *Int. J. Audiol.*, vol. 47, no. 6, pp. 356–361, Jan. 2008.
DOI: 10.1080/14992020801895136 47

[111] C. H. Taal, R. C. Hendriks, R. Heusdens, and J. Jensen, "An evaluation of objective measures for intelligibility prediction of time-frequency weighted noisy speech," *J. Acoust. Soc. Amer.*, vol. 130, pp. 3013–3027, Nov. 2011. DOI: 10.1121/1.3641373 47

[112] C. H. Taal, R. C. Hendriks, R. Heusdens, and J. Jensen, "An algorithm for intelligibility prediction of time-frequency weighted noisy speech," *IEEE Trans. Audio, Speech, Language Process.*, vol. 19, no. 7, pp. 2125–2136, Sep. 2011. DOI: 10.1109/TASL.2011.2114881 47, 48, 50, 56

Authors' Biographies

RICHARD C. HENDRIKS

Richard C. Hendriks Dr. ir. Richard C. Hendriks obtained his M.Sc. and Ph. D. degrees (both cum laude) in electrical engineering from Delft University of Technology, Delft, The Netherlands, in 2003 and 2008, respectively. From 2003 till 2007 he was a Ph.D. researcher at Delft University of Technology, Delft, The Netherlands. From 2007 till 2010 he was a postdoctoral researcher at Delft University of Technology. Since 2010 he is an assistant professor in the Signal and Information Processing Lab of the faculty of Electrical Engineering, Mathematics and Computer Science at Delft University of Technology. In the autumn of 2005, he was a Visiting Researcher at the Institute of Communication Acoustics, Ruhr- University Bochum, Bochum, Germany. From March 2008 till March 2009 he was a visiting researcher at Oticon A/S, Copenhagen, Denmark. His main research interests are digital speech and audio processing, including single-channel and multi-channel acoustical noise reduction, speech enhancement and intelligibility improvement.

TIMO GERKMANN

Timo Gerkmann Prof. Dr.-Ing. Timo Gerkmann studied electrical engineering at the universities of Bremen and Bochum, Germany. He received his Dipl.-Ing. degree in 2004 and his Dr.-Ing. degree in 2010 both at the Institute of Communication Acoustics (IKA) at the Ruhr-Universität Bochum, Bochum, Germany. From January 2005 to July 2005 he was with Siemens Corporate Research in Princeton, NJ, USA. In 2011 he was a postdoctoral researcher at the Sound and Image Processing Lab at the Royal Institute of Technology (KTH), Stockholm, Sweden. Since December 2011 he heads the Speech Signal Processing Group at the Universität Oldenburg, Oldenburg, Germany. His main research interests are on speech enhancement algorithms and modeling of speech signals.

JESPER JENSEN

Jesper Jensen Jesper Jensen received the M.Sc. degree in electrical engineering and the Ph.D. degree in signal processing from Aalborg University, Aalborg, Denmark, in 1996 and 2000, respectively. From 1996 to 2000, he was with the Center for Person Kommunikation (CPK), Aalborg University, as a Ph.D. student and Assistant Research Professor. From 2000 to 2007, he was a Post-Doctoral Researcher and Assistant Professor with Delft University of Technology, Delft, The Netherlands, and an External Associate Professor with Aalborg University. Currently, he is a Senior Researcher with Oticon A/S, Copenhagen, Denmark, where his main responsibility is scouting and development of new signal processing concepts for hearing aid applications. He is also a Professor with the Section for Multimedia Information and Signal Processing (MISP), Department of Electronic Systems at Aalborg University, Denmark. His main interests are in the area of acoustic signal processing, including signal retrieval from noisy observations, coding, speech and audio modification and synthesis, intelligibility enhancement of speech signals, signal processing for hearing aid applications, and perceptual aspects of signal processing.